About IFPRI

The International Food Policy Research Institute (IFPRI), established in 1975, provides research-based policy solutions to sustainably reduce poverty and end hunger and malnutrition. The Institute conducts research, communicates results, optimizes partnerships, and builds capacity to ensure sustainable food production, promote healthy food systems, improve markets and trade, transform agriculture, build resilience, and strengthen institutions and governance. Gender is considered in all of the Institute's work. IFPRI collaborates with partners around the world, including development implementers, public institutions, the private sector, and farmers' organizations. IFPRI is a member of the CGIAR Consortium.

About IFPRI's Peer Review Process

IFPRI books are policy-relevant publications based on original and innovative research conducted at IFPRI. All manuscripts submitted for publication as IFPRI books undergo an extensive review procedure that is managed by IFPRI's Publications Review Committee (PRC). Upon submission to the PRC, the manuscript is reviewed by a PRC member. Once the manuscript is considered ready for external review, the PRC submits it to at least two external reviewers who are chosen for their familiarity with the subject matter and the country setting. Upon receipt of these blind external peer reviews, the PRC provides the author with an editorial decision and, when necessary, instructions for revision based on the external reviews. The PRC reassesses the revised manuscript and makes a recommendation regarding publication to the director general of IFPRI. With the director general's approval, the manuscript enters the editorial and production phase to become an IFPRI book.

Genetically Modified Crops in Africa

Economic and Policy Lessons from Countries South of the Sahara

Edited by José Falck-Zepeda, Guillaume Gruère,
and Idah Sithole-Niang

A Peer-Reviewed Publication

International Food Policy Research Institute
Washington, DC

International Food Policy Research Institute
2033 K Street, NW
Washington, DC 20006-1002, USA
Telephone: +1-202-862-5600
www.ifpri.org

DOI: http://dx.doi.org/10.2499/9780896297951

Library of Congress Cataloging-in-Publication Data

Genetically modified crops in Africa : economic and policy lessons from
   countries south of the Sahara / edited by José Falck-Zepeda, Guillaume
   Gruère, and Idah Sithole-Niang.
      p.   cm.
   ISBN (10 digit): 0-89629-795-0
   ISBN (13 digit): 978-0-89629-795-1 (alk. paper)
   1. Transgenic plants—Africa, Sub-Saharan.   2. Transgenic plants—
Economic aspects—Africa, Sub-Saharan.   3. Crops—Genetic
engineering—Africa, Sub-Saharan.   4. Crops—Genetic engineering—
Economic aspects—Africa, Sub-Saharan.   I. Falck-Zepeda, José Benjamin.
II. Gruère, Guillaume P.   III. Sithole-Niang, Idah.   IV. International
Food Policy Research Institute.
SB123.57.G47815   2013
338.1'867—dc23   2013003854

Cover design: Carolyn Hallowell, Designer
Book layout: Princeton Editorial Associates Inc., Scottsdale, Arizona

Contents

Tables, Figures, and Boxes

Tables

Figures

Boxes

Abbreviations and Acronyms

AGBIOTECH	agricultural biotechnology
AGRA	Alliance for a Green Revolution in Africa
ARC	Agricultural Research Council
AU	African Union
BT	insect-resistant trait conferred on crops, allowing them to synthesize crystalline proteins that are lethal to specific insects (originates from the bacteria *Bacillus thuringiensis*)
CDO	Cotton Development Organisation
CERA	Center for Environmental Risk Assessment
CIMMYT	International Maize and Wheat Improvement Center
COMESA	Common Market for Eastern and Southern Africa
D&PL	Delta and Pineland
EAC	East African Community
EC	esophageal cancer
FAO	Food and Agriculture Organization of the United Nations
FAO-BIODEC	Food and Agricultural Organization of the United Nations online Database of Biotechnologies in Use in Developing Countries
FB_1	fumonisin B1
FY	fiscal year
GE	genetically engineered

GEF	Global Environment Facility
GM	genetically modified
GMO	genetically modified organism
HT	herbicide tolerant
IFPRI	International Food Policy Research Institute
IRMA	Insect-Resistant Maize for Africa
ISAAA	International Service for the Acquisition of Agri-biotech Applications
JECFA	Joint FAO/WHO Expert Committee on Food Additives
KAP	knowledge, attitudes, and perceptions
MISTIC	maximum incremental socially tolerable irreversible cost
MRC	Medical Research Council
NARO	National Agricultural Research Organisation (of Uganda)
NBC	National Biosafety Committee
NBF	National Biosafety Framework
NEPAD	New Partnership for Africa's Development
NTD	neural tube defect
PPP	public–private partnership
PROMEC	Programme on Mycotoxins and Experimental Carcinogenesis
R&D	research and development
SAGENE	South African Committee for Genetic Experimentation
SIRB	social incremental reversible benefits
SSA	Africa south of the Sahara
UNCST	Uganda National Council of Science and Technology
UNEP	United Nations Environment Programme
WEMA	Water-Efficient Maize for Africa
WHO	World Health Organization
WTP	willingness to pay
ZAR	South African rand

Foreword

The problems and constraints of finding sustainable solutions to end hunger and poverty in Africa are well documented. Agriculture is a critical sector in that quest, as it contributes approximately 35 percent of the continent's GDP while accounting for 70 percent of its labor force. The sector is also considered significant in the deployment of overall economic development strategies in the majority of African countries. While there has been some progress in improving African agriculture, productivity improvements are being overshadowed by new and in some cases increased challenges, such as population growth, a changing climate, and the fact that fewer than 30 percent of farmers have access to or use of improved seeds. In addition, technological innovations in use in other developing regions are proving difficult to institute or become accepted practice and policy in Africa—and policies for technological innovation will be essential to feeding Africa's population in the future.

One promising, yet controversial, technological innovation is biotechnology. Biotechnology tools, including genetically modified crops and other organisms, have produced valuable products that have been adopted by a large number of farmers globally, including those in Argentina, Brazil, China, and India. Genetically modified (GM) crops have been approved for commercial release in Burkina Faso, Egypt, and South Africa, while contained and confined testing of them has been performed in Kenya, Nigeria, Uganda, and other countries. Despite the documented benefits to farmers in developing countries, including those in Africa, many policymakers and farmers in Africa south of the Sahara remain hesitant about the use of GM crops, and they need more information about their potential, benefits, costs, and safety in the African context.

Calls to increase agricultural productivity, such as those found in the Maputo Declaration of 2003 and the statement from the African Union Summit of 2009, have not had an impact on the biotechnology debate, in spite of perceptions that biotechnology innovation has largely bypassed the region. While panels convened by the African Union to facilitate multi-stakeholder dialogue have recommended more policy research on economic, social, ethical, environmental, intellectual property, and trade issues relevant to Africa, only limited progress has been made.

The International Food Policy Research Institute organized the conference "Bringing Economic Analysis to Inform Biotechnology and Biosafety Policies in Africa" in Entebbe, Uganda, in May 2009 to take stock of economic and policy research focused on biotechnology in Africa and to identify relevant issues to help guide future research. This book brings together the papers written for, and presentations made at, that conference.

The agricultural productivity challenges facing Africa will require a pro-active response and vigilant evaluation of biotechnology and its many tools. *Genetically Modified Crops in Africa* is meant to help policymakers assess whether and how biotechnology can contribute sustainable solutions to ending hunger and poverty in Africa.

Shenggen Fan
Director General, IFPRI

Acknowledgments

The 2009 conference organized by the International Food Policy Research Institute (IFPRI) in Entebbe, Uganda, was made possible through support from the U.S. Agency for International Development (USAID).

We acknowledge the support of Dr. Theresa Sengooba and Ms. Christina Lakatos, who actively helped organize the conference. We thank the participants in the conference for their valuable contributions to the successful completion of the event.

We acknowledge further support from the International Development Research Centre–Canada (IDRC-Canada) and the Program for Biosafety Systems (PBS), which helped in the preparation of this book. PBS is funded by the Office of the Administrator, Bureau for Economic Growth, Agriculture and Trade/Environment and Science Policy, USAID, under the terms of award EEM-A-00-03-00001-00.

The opinions expressed herein are those of the co-editors and authors and do not necessarily reflect the views of IDRC-Canada or USAID or its missions worldwide.

Introduction and Background

José Falck-Zepeda, Guillaume Gruère, and Idah Sithole-Niang

Despite multiple internal and external investment efforts, Africa south of the Sahara (SSA) has not been completely successful in facing its agricultural and development constraints. This state of affairs has been complicated with the rise of increasingly complex constraints on the continent. The convergence of population growth, increased food production vulnerability, rising climatic variability, governance and political instability, and delayed investments to overcome environmental and agricultural productivity constraints appears to have thwarted agricultural development efforts. The region's vulnerability to these binding constraints on food security, as well as on economic growth and prosperity, can be seen in the fact that SSA is still enduring the impacts of the global food and financial crises that occurred in 2008 (IMF 2009; Arieff, Weiss, and Jones 2010; Brambila-Macias and Massa 2010).

Africa at the Crossroads

Indeed, the persistent impact of the global food and financial crises on SSA highlights the need for increased investments in the development of robust and resilient agricultural, food, fiber, and energy production systems. Increased investments focused on improving such systems can help address these countries' development challenges. Yet the historical record of investments in this area is littered with ambitious—and in many cases, failed—development plans, policies, and interventions. Easterly (2009) suggests that SSA focus on feasible, but homegrown, development interventions that seek solutions to specific problems.

Raising agricultural productivity is considered one of the most important ways to help develop a robust and resilient agriculture and to increase rural income (World Bank 2007). Many examples exist of successful interventions supporting agricultural development that have been built upon science, innovation, and the use of productivity-raising technologies (Spielman and Pandya-Lorch 2009). Among the broad set of available science and

1

technology interventions, genetically engineered (GE) crops[1] present an option that could help increase agricultural productivity, improve income, and contribute to achieving the goals of broader poverty alleviation and national development policies (FAO 2004).

The potential role of GE crops in addressing the continent's constraints has been recognized in Africa (Juma and Serageldin 2007). Accumulated experience and knowledge in Africa and other developing countries (see Qaim 2009; Smale et al. 2009; Pontifical Academy of Sciences 2010; Potrykus and Ammann 2010; Areal, Riesgo, and Rodriguez-Cerezo 2012) suggest that available GE crops in the short and medium term may have significant value for African agriculture. Yet their development and use remain controversial in many countries in SSA and in other developing countries. This is partly due to an incomplete understanding of the appropriate development role for GE crops and other biotechnologies that are products of nascent innovation systems in the subregion. Valuing the development potential for the introduction of GE crops in SSA must account for their potential development interventions within the scope of broader poverty alleviation efforts and national development policies, taking into account the economic and political contexts.

Like most technological interventions, GE crops will not solve all SSA development problems, nor will all available GE crops be useful—or appropriate—in the African context. Instead, African decisionmakers will need to evaluate the specific value of each GE crop as a tool in the portfolio of potential interventions that may be made available to farmers in the region. GE crops may be particularly important if they help solve specific crop productivity constraints in Africa. This is true especially of those productivity constraints that have not been resolved by conventional means, including conventional plant breeding, integrated pest management, and in those situations where other control/productivity enhancement approaches may not be accessible to farmers.

To identify potential beneficial technologies, an assessment of economic impacts is called for, which includes analyses at the farm, national, and international levels. In any such priority-setting exercise, the institutional setting needs to be accounted for, as it may have an impact on adoption, technology use, and output marketing by farmers in developing countries (see, for example,

---

1 Here we use the label "GE" for transgenic crops or products derived thereof, because it is one of the most commonly employed term used in the literature and public media. Other equally imprecise terms (such as genetically modified (GM) crops derived from modern biotechnology) could be used instead, and may be used interchangeably in the book. In this book we focus on GE crops, yet many of the issues discussed here apply to animals, arthropods and other insects, and microorganisms, which may become available in SSA.

Tripp 2009). So far, there are only limited examples of GE crop impact assessment. Smale et al. (2009) review the economic literature assessing GE crops in African and other developing countries' agriculture, including their impact on farmers, households, communities, and trade, and the institutional context in which these technologies may be deployed for potential use by farmers. They draw the four following lessons.

- On average, adoption has been profitable to users—but averages mask variability in agroclimates, host cultivars, and farming practices.

- There are too few traits under study, and too few cases and authors—generalizations should not be drawn. More time is needed to describe the effects of adoption.

- During the next decade, practitioners will need to address cross-cutting issues for further study, such as impacts on poverty, gender, and public health.

- To address broader issues, impact assessment practitioners need to develop improved methods and draw from multidisciplinary collaborations.

Similar lessons have been reported in Qaim (2009, 2010); in recent meta-analyses of impact assessment studies of insect-resistant cotton by Finger et al. (2011); and for all GE crops in Areal, Riesgo, and Rodriguez-Cerezo et al. (2012).

The relatively limited research on the impact of GE crops implies that its contribution to the policy debate on GE crops and biotechnology in the continent has also been limited. Moreover, the policy debate is being undermined because much of the discussion in Africa has yielded to external (and in some cases to internal) pressures to move away from science and rational debate and discussion, toward either antagonistic or unconditionally supportive views on GE crops (Novy et al. 2011; Takeshima and Gruère 2011). These polarized views generally lack robust data and evidence-based policy analysis, contributing to confusion about the real or potential value of GE crops for Africa's agriculture, especially in African policy debates.

This book is an attempt to move the discussion away from polarized positions. It aims to contribute to a rational debate on the actual benefits, costs, and risks of existing and future GE crops and technologies for Africa. To accomplish this goal, we introduce a broad set of contributions documenting issues relevant to the current African policy debate. These contributions are representative of the state-of-the-art knowledge about GE biotechnologies in

Africa. We also include references to other papers and materials relevant to the debate when appropriate, which may help elucidate important questions for the proper assessment of GE crops and similar technologies in Africa.

The following sections aim at setting the stage for the policy debate on GE crops. We briefly present the status of GE crop adoption and capacity in Africa. We then list a number of key potential constraints and describe some of the internal positions on biotechnology.

GE Crops: "Miracle Crops" or "Frankenfoods"?

GE crops have been portrayed unequivocally by those opposing or promoting the technology as either *the* solution for feeding the world, or in some instances, as crops that would bring environmental and social catastrophes of incalculable consequences (Brac de la PerriFre and Seuret 2000; GRAIN 2004). Stone (2002) documents that such contrasting positions can and do obscure many of the complexities involved in GE crop adoption and their use in developing countries. A more balanced position would consider GE crops neither as Frankenfoods nor as Miracle Crops per se, but rather as a set of technologies with unique attributes, different from past innovations, such as the Green Revolution's maize and wheat varieties. In particular, many of these technologies have been produced using research and development (R&D) inputs that are protected intellectual property, and most technologies now available commercially have been developed by the private sector.

These attributes need to be characterized, discussed, and in some cases addressed to ensure their compatibility with and support of poverty alleviation efforts, especially in the agricultural context of SSA. Furthermore, there is a need to clearly separate the intrinsic production, productivity, and socioeconomic impacts of GE crops from more general concerns expressed by some stakeholders over forced industrialization, corporate control of agriculture, and the impact of technology on traditional farmers and agricultural practices and communities. The latter implies the need to conduct in-depth social and institutional context analysis in which these varieties may be or have been released to farmers to ensure that society can determine the potential role of GE crops in development and poverty alleviation efforts (Stone 2011). This is especially important in the African context.

Stakeholder Support for GE Crops

GE crop supporters present these technologies as a distinct option to promote food security and sustainable agriculture for developing countries. GE crops open the possibility of addressing biotic and abiotic constraints to food, feed, and fiber production. GE crops may, for example, enhance productivity, improve pest and weed control, and increase tolerance to drought and salinity. These crops may also improve public health through reductions in pesticide applications and through enhanced nutrition, such as vitamin A–enhanced rice that is currently being evaluated in a number of developing countries. Yet different stakeholders have contrasting positions toward these technologies.

Stakeholder Criticisms of GE Crops

Some stakeholders often cite the fact that existing GE crops were largely developed by the private sector for use in industrialized countries in an intensive and commercially focused agriculture. The consequence of this approach, in their view, is that existing GE crops are inappropriate for traditional agriculture as practiced in Africa and other developing countries. They believe that this approach empowers private firms to exercise monopoly power and thus price the technology at a higher level than in a competitive market (Moschini and Lapan 1997; Falck-Zepeda, Traxler, and Nelson 2000). Private sector–led agricultural R&D is a different pathway than that taken by previous agricultural innovation processes, which have been driven mostly by the public sector. The private sector–led R&D investments and continued control over GE crops is seen by some commentators as one more example of corporate control of agriculture and its activities.

Other issues have been raised, such as the "contamination" of traditional varieties due to pollen flow, uncontrolled gene dispersion, impacts on trade, disruption of traditional communities and livelihoods, dependency on private sector, production risk increasing due to the rise of monocultures, and the decline of smallholder crop diversification.[2] These concerns may or may not be unique to GE crops. They may also belong to a larger set of general concerns over the role of science and technology in contributing to poverty alleviation and development. Furthermore, it is not always clear whether these

---

2 For a summary of the biological and environmental issues, see Conner, Glare, and Nap (2003). For a broader discussion that also includes social issues, see Uzogara (2000) and Stone (2002). In some cases and under a relatively complex set of conditions, the introduction of modern varieties—including GE crops—can introduce the potential for private firms exercising monopoly power over resource-poor households and farmers in developing countries. This outcome is valid but dependent on a set of conditions that determine its likelihood. In particular, it is a possible scenario where market conditions are such that they force farmers in developing countries to become members of captive markets with little or no choice for a diversity of crop varieties or other production alternatives. Therefore conventional or traditional varieties preserved by farmers could disappear over time (Munro 2003; Knezevic 2007).

concerns apply to existing technologies or to proposed technologies in the development pipeline for the African context.

Toward a More Balanced Approach

Certainly, some of these stakeholder concerns and issues may be valid for some crops or incorporated traits through genetic engineering in some locations. In our view, each concern is an empirical issue that has to be identified and analyzed as part of an ex ante assessment of GE crops before deliberate release into the environment.

A prudent approach would consider the relevant facts and then render a robust and complete analysis of the appropriateness of a specific technology for its intended target country or region. One cannot generalize that GE crops are either unequivocally Frankenfoods or Miracle Crops. A rational approach would require judging GE crops on a case-by-case basis while considering all the costs, benefits, and risks estimated through robust assessments.

Still, following Norman Borlaug's and Jimmy Carter's opinions (Paarlberg 2008), we believe that the GE crop assessment process, as it stands now, needs to concentrate less on their potential risk (especially in the case of well-studied technologies where there is an established record of safety and use in other countries) and more on their actual impact and on their access by poor farmers. A redirected focus on development efforts is needed to ensure that poor farmers could benefit from the assessed, safe, relevant, and beneficial GE crops in their context. Separating generalized issues from the facts germane to useful GE crop development, deployment, access, and performance is one of the motivating factors behind this book.

GE Crop Adoption: The Reality behind the Numbers

The area planted to GE crops has increased at a rapid pace since the release of the first commercial crops in the United States and China. As of 2011, there were approximately 160 million hectares worldwide cultivated with the crops (James 2011). Yet the growth in area planted continues to be focused mainly on four crops (soybeans, maize, cotton, and canola) and two traits (herbicide tolerance and insect resistance). Small areas in several countries have been planted with other crops, including alfalfa, beans, tomatoes, petunias, papayas, potatoes, sweet peppers, squash, carnations, and sugar beets.

Developing countries cultivate approximately 50 percent of the global area devoted to GE crops, totaling close to 80 million hectares. Importantly,

FIGURE I.1   Area cultivated with genetically engineered crops, by country

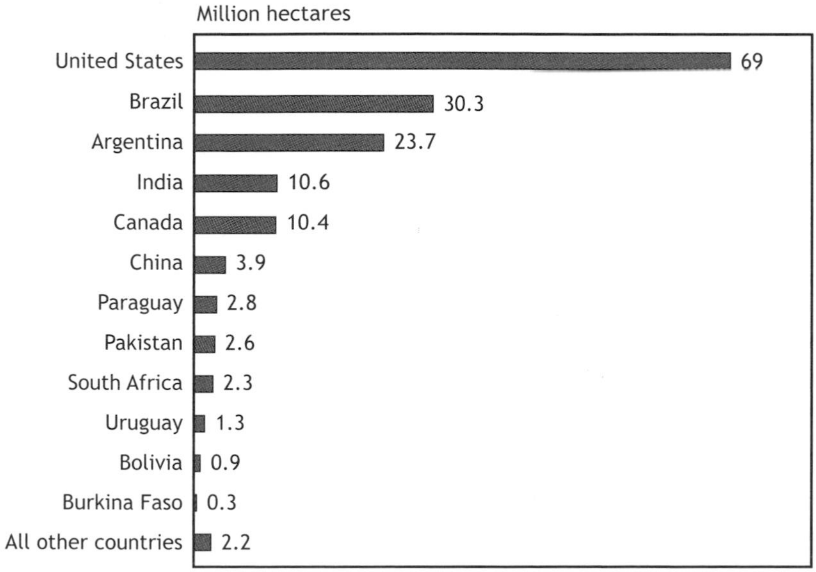

Million hectares

Country	Million hectares
United States	69
Brazil	30.3
Argentina	23.7
India	10.6
Canada	10.4
China	3.9
Paraguay	2.8
Pakistan	2.6
South Africa	2.3
Uruguay	1.3
Bolivia	0.9
Burkina Faso	0.3
All other countries	2.2

Source: Data extracted by Patricia Zambrano and Jose Falck-Zepeda from James (2011).

the share planted by developing countries has been increasing over time and may even become higher than that of developed countries (James 2011). As shown in Figure I.1, developing countries with the highest area planted in 2011 included Brazil (30.3 million hectares), Argentina (23.7 million), India (10.6 million), China (3.9 million), Paraguay (2.8 million), and Pakistan (2.6 million).

In 2011, the share of GE crops planted in Africa was small. Africa's share was less than 1.6 percent of the total area planted to GE crops (see Figure I.2). Similar to the evolution across all developing countries, the area planted by African countries has been increasing over time. Furthermore, the number of crops in the product development and the biosafety regulatory pipeline and those that have been commercialized have also been increasing over time in the continent. The need then exists to examine the current status of GE crops in the product development and regulatory pipelines as a necessary background to understand the issues discussed in this book. The next section addresses these issues in some detail.

FIGURE I.2   Share of genetically modified crop acreage, by country

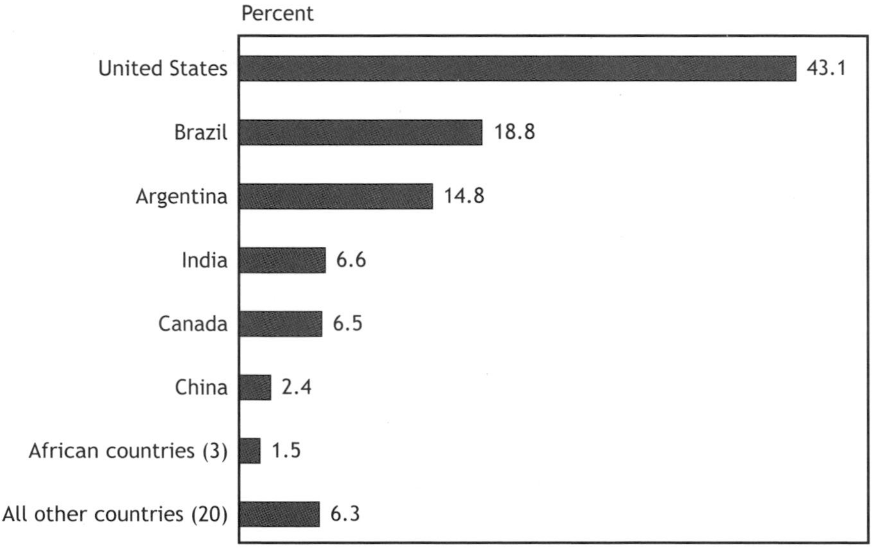

Percent

Country	Percent
United States	43.1
Brazil	18.8
Argentina	14.8
India	6.6
Canada	6.5
China	2.4
African countries (3)	1.5
All other countries (20)	6.3

Source: Extracted by Patricia Zambrano from James (2011).

Africa's Biotechnology Capacity, Biosafety Status, and Adoption Impact

The status of biosafety regulatory frameworks and of GE crop technologies play a role in the observed low level of GE crop adoption in Africa. Currently, there are no completely indigenous GE crop technologies generated in Africa that we are aware of, except possibly in Egypt. Most technologies under development for SSA use genetic constructs or transformation procedures developed elsewhere, and plant germplasm materials are usually selected and developed internally in the country of interest.

Biosafety Regulatory Framework Status

The foundation of biosafety regulatory systems, the national biosafety framework (NBF), includes policies, laws, and implementation regulations. Countries implement their national biosafety framework by mobilizing human, financial, and technical capacities in the country, which permits agents to conduct a biosafety assessment and then submit a recommendation, or in some cases, a decision.

Figure I.3 presents a map of the status of NBFs in Africa as of 2009. Their status in this figure is described as functional, interim, work in

FIGURE I.3   Status of National Biosafety Frameworks (NBFs) in Africa

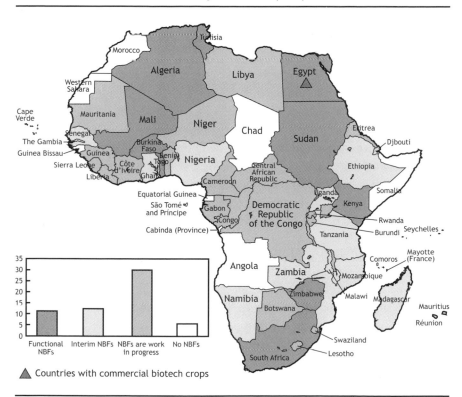

Source: Karembu, Nguthi, and Ismail (2009).

progress, and nonexistent, depending on the degree of evolution shown by countries at that particular point in time. It should be noted that in this figure, even if an NBF has been described as "functional," it does not necessarily mean that the country has approved a GE crop for deliberate release. Country statuses have not changed much since this figure was developed. There were two relevant developments: Uganda has approved several confined field trials and is developing a formal policy, and Nigeria has approved a policy and is in the process of developing implementation regulations and approving confined field trials.

At the same time, the biosafety regulatory development process has evolved sufficiently to permit confined field trials in Burkina Faso, Egypt, Kenya, Nigeria, Uganda, and Zimbabwe, and commercialization release in Burkina Faso, Egypt, and South Africa. The ability to conduct confined field trials is a parameter that helps document biosafety assessment capacity as it is usually, but not always, a first step before the process of assessment for commercial release. Applications for confined field trials are pending in many more countries.

Biotechnology Product Development/Regulatory Pipeline

Table I.1 lists GE crop technologies under product development or biosafety regulatory assessment. The number of GE crops is expanding beyond the four crops with two traits planted as the dominant share of global GE crop area; the number of countries testing and approving them is also expanding. A noteworthy development is the rise of crops of special interest to some countries in Africa, including bananas, sweet potatoes, cowpeas, and cassava. Furthermore, the number of traits in development is also increasing. These include resistance to fungal and viral diseases and tolerance to drought conditions.

GE Cotton and Maize Adoption and Socioeconomic Impact in Africa

According to the estimates of James (2011), in 2011, South Africa cultivated approximately 2.3 million hectares, Burkina Faso 0.3 million hectares, and Egypt less than 0.05 million hectares of GE crops. Maize and cotton account for most of the area planted to GE crops in Africa.

More specifically, South Africa planted 2.3 million hectares of GE maize, cotton, and soybeans. Burkina Faso cultivated 300,000 hectares of insect-protected cotton (Bt) representing approximately 70 percent of total cotton area, which represents a significant increase from 2009, when roughly 25 percent of the area was planted to Bt cotton (James 2011). In Egypt, the actual area planted has been quite small since the commercialization approval granted by the competent authority in 2008. Much of the area planted has been for seed reproduction purposes, although there are some reports that the expectation in Egypt is that area planted to GE maize will increase rapidly in the near future (Adenle 2011; James 2011).

The socioeconomic impact assessment literature on the effects of GE crops in SSA is relatively thin. As shown in Smale et al. (2009), much of the ex post assessment work has been done on measuring producer impacts from the adoption of GE crops, mostly focused on insect-resistant cotton in South Africa, with a handful of reports from Burkina Faso. There is a small, but growing, ex ante assessment body of literature for proposed technologies in Africa. Reports from South Africa and other developing countries that have adopted GE crops show that considerable spatial, temporal, and user variabilities exist and have an impact on results. These conclusions have been validated in a formal meta-analysis done by Finger et al. (2011).

TABLE I.1   Regulatory status of genetically engineered crops in the regulatory and development pipeline, 2009

Country	Crop	Trait	Genetic event	Institution	Regulatory status
Kenya	Maize (*Zea mays* L.)	Insect resistance	Mon 810, Cry1Ab 216, Cry1Ba	KARI, CIMMYT, Monsanto, University of Ottawa, Syngenta Foundation, Rockefeller Foundation	Confined field trials
	Cotton (*Gossypium hirsutum* L.)	Insect resistance	Bollgard II	KARI, Monsanto	Confined field trials
	Cassava (*Manihot esculenta*)	Cassava mosaic disease resistance	AC1-B	KARI, Danforth Plant Science Center	Confined field trials
	Sweet potato (*Ipomoea batatas*)	Viral disease resistance	CPT 560	KARI, Monsanto	Confined field trials
Uganda	Cotton (*Gossypium barbadense*)	Insect resistance, herbicide tolerance	Bollgard IR/HT	NARO, Monsanto, ABSPII, USAID, Cornell University	Confined field trials approved
	Banana (*Musa* sp.)	Black sigatoka resistance	Chitinase gene	NARO, University of Leuven	Confined field trials
				IITA, USAID	Confined field trials
	Cassava (*Manihot esculenta*)	CMD and cassava brown streak disease (CBSD)		NaCRRI, International Potato Center, Danforth Plant Science Center	Application for confined field trials approved by the NBC
Nigeria, Burkina Faso, Ghana	Cowpea (*Vigna unguiculata*)	Insect resistance	Cry1Ab and nptII genes	AATF, NGICA, IITA, Purdue University, Monsanto, Rockefeller Foundation, USAID, Department for International Development, CSIRO, Institut de l'Environnement et de Recherches agricoles (Burkina Faso), Institute of Agricultural Research (Ghana), Kirkhouse Trust	Confined field trials approved in Nigeria
Kenya, Tanzania, Uganda, South Africa, Mozambique	Maize (*Zea mays* L.)	Drought tolerance	CspB-Zm event 1	AATF, National Agricultural Research Institutes in the five countries, CIMMYT, Monsanto, Bill & Melinda Gates Foundation, Howard G. Buffett Foundation	Confined field trials pending regulatory approval in Kenya; confined field trials in South Africa ongoing

TABLE I.1   (continued)

Country	Crop	Trait	Genetic event	Institution	Regulatory status
South Africa, Burkina Faso, Kenya	Sorghum (*Sorghum bicolor*)	Nutrition enhancement		Consortium of nine institutions led by the Africa Harvest Biotech Foundation International and funded by the Bill & Melinda Gates Foundation	Contained greenhouse trials in Kenya and South Africa
South Africa	Maize (*Zea mays* L.)	Drought tolerance	MON 89034, MON 87460	Monsanto	Confined field trials
		Herbicide tolerance	Syngenta GA21	Syngenta	Field trial release
		Insect resistance	Syngenta MIR162		Field trial release
		Insect/herbicide tolerance	Syngenta BT11 × GA21		Field trial release
			BT11 × MIR162		Field trial release
			Pioneer 98140	Pioneer	Confined field trials
			Pioneer 98140 × Mon 810	Pioneer	Confined field trials
	Cassava (*Manihot esculenta*)	Starch enhancement	TMS60444	Agricultural Research Council in South Africa–Institute for Industrial Crops	Contained trial
	Cotton (*Gossypium hirsutum* L.)	Insect/herbicide tolerance	Bayer BG11 × RR FLEX	Bayer	Trial release
			GHB119		Trial release
			BG11 × LLCotton25		Trial release
			CottonT304-40		Trial release
		Herbicide tolerance	CottonGHB614		Trial release
			CottonGHB614 × LLCotton25		Trial release
	Potato (*Solanum tuberosum* L.)	Insect resistance	G2 Spunta	Agricultural Research Council in South Africa–Onderstepoort Veterinary Institute	Field trials
	Sugarcane (*Saccharum officinarum*)	Alternative sugar	NCo310	South African Sugarcane Research Institute	Field trials

TABLE I.1   (continued)

Country	Crop	Trait	Genetic event	Institution	Regulatory status
Egypt	Maize (*Zea mays* L.)	Insect resistance	Mon 810	Monsanto	Approved for commercialization
			n.a.	Pioneer	Field trials
	Cotton (*Gossypium barbadense*)	Salt tolerance	MTLd	Agricultural Genetic Engineering Research Institute	Contained greenhouse trials
	Wheat (*Triticum durum* L.)	Drought tolerance	HVA1	Agricultural Genetic Engineering Research Institute	Field trials
		Fungal resistance	Chitinase		Contained greenhouse trials
		Salt tolerance	MTLd		Contained trial
	Potato (*Solanum tuberosum* L.)	Viral resistance	Cry V	Agricultural Genetic Engineering Research Institute	Field trials
		Viral resistance	CP-PVY		Field trials
	Banana (*Musa* sp.)	Viral resistance	CP-Banana CMV		Contained trial
	Cucumber (*Cucumis sativus*)	Viral resistance	Cp-ZYMV		Field trial
	Melon (*Cucumis melo*)	Viral resistance	Cp-ZYMV		Field trial
	Squash (*Cucurbita pepo*)	Viral resistance	Cp-ZYMV		Contained trial
	Tomato (*Lycopersicon esculentum*)	Viral resistance	CP- REP-TYLCV		Contained trials

Source: Karembu, Nguthi, and Ismail (2009).

Note: AATF = African Agricultural Technology Foundation; ABSPII = Agricultural Biotechnology Support Project II; CIMMYT = International Maize and Wheat Improvement Center; CSIRO = Commonwealth Scientific and Industrial Research Organisation; IITA = International Institute of Tropical Agriculture; KARI = Kenyan Agricultural Research Institute; n.a. = not available; NaCRRI = National Crops Resources Research Institute; NARO = National Agricultural Research Organisation; NGICA = Network for the Genetic Improvement of Cowpea for Africa; USAID = United States Agency for International Development.

From Concept to Farmers:
Issues for GE Crops in the African Context

This rapid review of the situation in SSA triggers a number of questions. Why aren't more GE crops being deployed in Africa? Should we be seeing more GE crops being deployed in Africa? What are the appropriate crops and traits that may support smallholder agriculture in Africa? Will these technologies contribute to poverty alleviation efforts?

The answers proposed to these questions vary significantly in the literature. Current and future GE crops clearly face a number of deployment challenges. Some challenges are common to the dissemination of all new technologies in SSA. Other challenges are unique to GE crops and may require innovative approaches for addressing these constraints in a meaningful way. In the next section, some of these issues are discussed in a technology framework chain with the intention of properly situating the issues, constraints, benefits, and other relevant issues to a decisionmaker in Africa.

Technology Development, Adaptation, and Dissemination Issues

SSA faces a situation where there is a growing but still insufficient level of investment in R&D, especially in a select group of countries (Beintema and Stads 2010). The generalized level of R&D investments in SSA translates into a relatively poor biotechnology innovative capacity. This in turn affects the capacity for conducting GE adaptive and targeted R&D in the continent. The low availability of human and financial resources in the region limits the overall innovative capacity to conduct R&D and to develop indigenous GE crops. Unlike other crops, GE crops have to comply with biosafety regulations and be subject to risk assessments. To do so, countries have to sustain a sufficient decisionmaking capacity. Although many SSA countries are still lacking a regulatory framework or a scientific and regulatory capacity to implement such regulations, some notable efforts are underway to develop robust biosafety systems.

National innovative capacity that can develop those GE traits of interest to national priorities has to be weighed against the possibility of accessing such technologies developed elsewhere. From a science and technology standpoint, African decisionmakers can opt for different innovative capacity systems to deliver products to farmers. Some countries in Africa, though, lack even the minimal investments in R&D capacity necessary to conduct adaptive R&D in their own national research systems. Furthermore, some countries in the region have expressed a concern over the potential risks of these technologies to their farmers. Such concerns are not insurmountable hurdles, as there are practical and feasible approaches that can empower countries to ensure an acceptable level of safety after a scientific assessment, while ensuring innovation and technology transfer to their farmers.

African seed systems and germplasm delivery mechanisms face a number of constraints that have been described in the literature. Technology delivery and dissemination are limited by institutional weaknesses and the insufficient development of national seed systems in Africa. These constraints are not unique to GE crop seeds. Yet they can be magnified as the need arises to ensure a working system that facilitates knowledge and information exchange about the use of the technology, and market signal transmissions, including price premiums or value-added paid for competing commodities in markets.

Developing an appropriate science and technology strategy that directs investments in R&D capacity and strategy toward identified GE crops is critical. To do so, the costs, benefits, and risks that GE technologies may pose to farmers need to be thoroughly examined, in order to select "best bet" strategies to address specific productivity constraints in Africa.

Obstacles Related to Adoption

Farmers considering the adoption of GE crops face many institutional challenges. Access to credit and complementary inputs, their ability to manage production risk, and other binding institutional constraints play determinant roles in their decisions. A very important issue identified in a growing number of studies is farmers' access to knowledge and information about the use of the technology and its market potential. Many of the GE crops available for adoption in and outside Africa have demonstrated their technical capability to provide benefits to farmers. However, a farmer's ability to tap into those potential benefits can be limited by institutional issues.

This calls for policymakers to consider supporting the policy and institutional environment to maximize the benefits and minimize the risks of GE crop adoption in Africa. In this sense, the policy environment will strive to avoid potential cases of "technological triumphs but institutional failures" observed (Gouse et al. 2005, 1).

Marketing/Trade in Local and External Markets and Related Issues of Consumer Acceptance

The introduction and use of a GE crop in any country can potentially result in loss in export markets to trade-sensitive countries for the specific crop being considered. In some situations, African countries have argued that the potential approval and use of GE crops can even lead to the potential loss to unrelated export markets. Although such perceptions may be misplaced, African countries may face pressures by private buyers and consumer demand in trade-sensitive countries (Gruère and Sengupta 2009; Gruère and Takeshima 2012).

Market risks associated with potential external trade losses due to the adoption of GE crops are magnified by the growing trend in consumer concerns in African countries, especially among urban consumers, and by labeling and related marketing requirements in some consumer-sensitive countries around the world, especially Europe, some countries in Asia, and the Middle East.

An additional issue is the increased possible trade losses associated with asynchronous or asymmetric approvals of GE crops. GE crops that are approved in one country but not in other trading partners can result in significant trade disruptions. Because borders in Africa can be porous, and as regional trade increases within Africa, this issue may grow in importance. This calls for policies examining potential export losses with trade-sensitive countries and across regions in Africa to search for an adequate management strategy.

The trend toward a regionalization of biosafety assessment procedures and the capacity to establish a regulatory decisionmaking process also require support. Efforts such as those in the Common Market for Eastern and Southern Africa (COMESA) and the West African Economic and Monetary Union, which seek regional approaches to biosafety assessments and in some cases to decisionmaking, will become more important with increasing trade activity at the subregional level.

Contrasting Stakeholder Positions in Africa

Although the problems can be identified, the complexity of the African political debate around GE products has to be accounted for. In particular, there are many contrasting positions with regard to GE crops in SSA. For example, the High-Level Panel on Biotechnology report commissioned by the African Union (AU) and the New Partnership for Africa's Development (NEPAD) states that

> Africa needs to take strategic measures aimed at promoting the application of modern biotechnology to regional economic integration and trade. Such measures include fostering the emergence of regional innovation systems in which biotechnology-related Local Innovation Areas play a key role. [Juma and Serageldin 2007, xix]

The AU/NEPAD Panel laid out some of the preconditions for taking advantage of GE crops to support economic development efforts, especially with regard to regional economic coordination efforts to ensure proper and efficient regulatory assessment processes. Suggested by the AU/NEPAD Panel, a regional approach to biotechnology and biosafety regulations is being pursued independently by COMESA and by the Economic Community of West African States. These efforts are examining different modalities for developing regional approaches to biosafety regulations that seek to address risk-assessment procedures combined with national or regional decisionmaking. They do not represent an explicit endorsement of biotechnology or GE crops per se; rather, they open the possibility of examining these technologies on a case-by-case basis.

In turn, in its official "Statement on Plant Breeding and Genetic Engineering," the Alliance for a Green Revolution in Africa (AGRA) has this position with regard to GE crops and other organisms:

> Our mission is not to advocate for or against the use of genetic engineering. We believe it is up to governments, in partnership with their citizens, to use the best knowledge available to put in place policies and regulations that will guide the safe development and acceptable use of new technologies, as several African countries are in the process of doing. We will consider funding the development and deployment of such new technologies only after African governments have endorsed and provided for their safe use. Our mission is to use the wide variety of tools and techniques available now to make a dramatic difference for Africa's smallholder farmers as quickly as possible. [AGRA 2010]

AGRA's position is basically neutral; it leaves any decision in terms of future investments pending on decisions taken at the national level. This decision contrasts with national developments. Burkina Faso, Egypt, and South Africa have allowed the commercial cultivation of these crops by authorizing their deliberate release into the environment. More countries have approved confined field trials or have invested in the development of GE crops developed by the public sector for those issues of interest to African countries. This position also contrasts with that of countries that have explicit restrictive policies with regard to the importation of GE foods ranging from a complete ban on all imports (Zambia) to a ban on imports unless processed or milled (Angola, Lesotho, Malawi, and Zimbabwe).

Positions for or against the technology are not limited to the national level. Some organizations with stated missions that include social justice, biodiversity protection, small farmer livelihood assurance, and food security (such as the Third World Network, Via Campesina, Greenpeace, Oxfam, GRAIN, or

Friends of the Earth International) have argued and conducted campaigns against the deployment of GE technologies in developing countries, including those in Africa. It is not always clear whether these organizations' positions are against GE crops per se, an opposition to existing GE crops being deployed in developing countries, or a reaction to industrialization, multinational corporation development, or privatization of agricultural research, but they are vocally opposed to GE crops.

Although some of these groups continue to push their own agendas, they may not have addressed the complex issues surrounding the potential introduction and dissemination of GE crops, including consideration of the potential benefits from adoption and the reality that not adopting a GE crop could also have consequences (Pew Initiative on Biotechnology 2004). Although there may be some risks associated with the use of GE crops, the status quo in terms of conventional crops is not riskless either. The influence of donor and partner countries plays a key role in this area. So do trade relationships and various international pressure groups.

In fact, decisionmakers in SSA are bombarded with multiple, mostly conflicting positions and messages about the appropriateness of GE crops in the African context. The policy debate milieu grew to such a chaotic state that the AU's declaration after its 2006 Ministerial Meeting included text that described the situation poignantly:

> The two extreme positions have tended to confuse many African policymakers and sections of the public because of the lack of reliable information and guidance available to these groups. There is uncertainty and confusion in many of the African governments' responses to a wide range of social, ethical, environmental, trade and economic issues associated with the development and application of modern genetic engineering. The absence of an African consensus and strategic approaches to address these emerging biotechnology issues has allowed different interest groups to exploit uncertainty in policymaking, regardless of what may be the objective situation for Africa. [African Union 2006, 1]

In this setting, policy research has a role to play in examining the potential and actual use of GE crops and related issues to its deployment. As an important African policymaker said at the 5th Conference of Parties of the Biosafety Protocol in Nagoya, Japan:

> Given the lack of consensus amongst countries and the conflictive context, it is therefore imperative that any GE crop assessment work be

buttressed by proper science (natural or social), otherwise it will be crit-
icized by those opposing or promoting this technology. It is a mammoth
task, yes, but somebody has to start the ball rolling. It's a challenge that
we must embrace. [A. Mafa, pers. comm., 2010]

This book is a first step in this direction. In this complex and competitive
setting, information is critical for multiple purposes. When writing a policy,
drafting regulations, or making discrete decisions on GE crops, policymakers
in SSA have to use selected information to advance their goals. Yet credible
objective information on the impacts of GE crops and products based on
their cost, risks, and benefits is not always easy to find and to digest, given
the complexity of some of the issues at stake. Furthermore, even policy analysts, researchers, and academics involved in agriculture policy may find it
hard to find, access, and synthesize peer-reviewed studies on GE crops in the
African context. This is particularly true when gauging the economic effects of
GE crops.

Why This Book?

Several countries in SSA have expressed concerns related to the farm-
level impacts, consumer concerns, trade impacts, and the biosafety and other
regulatory issues related to GE crops and are considering inclusion of socio-
economic assessments in their technology approval processes.[3]

For example, discussions during the approval process and subsequent adoption of GE crops in Burkina Faso and South Africa have generated significant
internal controversies related to the potential socioeconomic, institutional,
political, and environmental effects of technology adoption. Controversies and
sometimes acrimonious discussions included socioeconomic concerns about
the potential market effects of local adoption of Bt cotton, impacts on resource-
poor farmers, farmers' dependence on a continuous flow of innovations, as
well as external impacts that may affect local farmers (such as the potentially
adverse reaction in some European markets). Other important concerns raised
by opponents of the technology are the potential environmental and ecological implications of GM technologies, all of which bring additional uncertainty
to the likelihood of farmer adoption. Examples of these discussions and debates
include Pschorn-Strauss (2005) and Moola and Munnik (2007).

---

3 Example of countries considering such policies can be found in Mulenga and Shumba-Mnyulwa
(2010) and Falck-Zepeda (2009).

These questions have also been raised in national and international forums in the context of biosafety regulatory and technology decisionmaking processes. African decisionmakers and all stakeholders involved in the process raise these questions, as they aim at identifying potential interventions to address specific productivity issues that impinge directly on farmer livelihoods.

A better understanding of the development, delivery, and downstream impact of GE crop innovations is required to comprehend their potential role in the African context. This will ensure that the right crops, traits, and delivery methods are identified and used. Furthermore, understanding knowledge processes and the institutional framework in which GE crops may be deployed can help ensure maximization of the potential benefits while minimizing the risk to African farmers and communities.

The overall objective of this book is to contribute to reducing the knowledge gap about the potential role and impact of GE crops in SSA. The volume gathers a set of policy and economic studies recently completed on the current or potential effects of GE products in the countries of this region. Although the collection does not claim to be exhaustive in any way, it provides a discussion of relevant issues discussed in SSA and other policy forums, as well as some new and emerging themes. This book addresses some of the key policy questions in the debate on the role of GE crops in the region. The targeted audience includes policy analysts, policymakers, scientists, researchers, university students, and other stakeholders working on policy issues related to agricultural biotechnology in Africa and who are interested in an accessible volume on policy analysis.

The collection of studies is based on updated contributions that were initially presented at a conference organized by the International Food Policy Research Institute (IFPRI) in Entebbe, Uganda, in May 2009. Chapters are organized thematically in three parts. Part I consists of three chapters on the economic effects of GE crops, with a focus on specific technologies. Part II presents two chapters on market acceptance, including one on consumer acceptance of GE food in Uganda and another discussing potential trade risk and regional integration. Part III focuses on research, regulatory, and technology delivery issues. Although each chapter addresses one specific question, it also provides general lessons for policymakers. The book ends with a conclusion section that collects lessons and issues for policy and decisionmaking and identifies areas for future research.

Throughout the book, care has been taken to consider the distinct opinions and positions in the debate. IFPRI's policy toward biotechnology is that even though some of these technologies are controversial and alone cannot solve complex poverty and food insecurity issues, some of them have the potential to address specific issues related to hunger and malnutrition in developing countries. Because of these considerations

> IFPRI believes it would be irresponsible not to assess the potential of genetically modified crops such as nutrient-enriched or drought-tolerant and disease-resistant crop varieties. At the same time, the Institute fully supports appropriate biosafety regulatory systems that are able to assess the risks. [IFPRI 2013]

We hope that this contribution will help inform the debates, in Africa and elsewhere, about the current and potential economic role of these crops in the agriculture of SSA. Furthermore, we expect that this book will help identify current knowledge gaps and engage the innovation, product delivery, and downstream impacts related to GE crops in Africa in a more systematic manner, while at the same time providing relevant and timely information to the ongoing discussions related to the potential adoption and use of GE crops.

References

Adenle, A. A. 2011. "Adoption of Commercial Biotech Crops in Africa." *International Proceedings of Chemical, Biological and Environmental Engineering* 7: 176–180. Accessed January 31, 2013. http://www.ipcbee.com/vol7.htm.

African Union. 2006. "An African Position on Genetically Modified Organisms in Agriculture." Accessed November 30, 2010. www.africa-union.org/root/AU/AUC/Departments/HRST/biosafety/DOC/level2/AfricanPositionOnGMOs_EN.pdf.

AGRA (Alliance for a Green Revolution in Africa). 2010. "Statement on Plant Breeding and Genetic Engineering." Accessed December 10, 2010. www.agra-alliance.org/section/about/genetic_engineering.

Areal, F. J., L. Riesgo, and E. Rodriguez-Cerezo. 2012. "Economic and Agronomic Impact of Commercialized GM Crops: A Meta-Analysis." *Journal of Agricultural Science*. Available on CJO2012 doi:10.1017/S0021859612000111.

Arieff, A., M. A. Weiss, and V. C. Jones. 2010. "The Global Economic Crisis: Impact on Sub-Saharan Africa and Global Policy Responses." U.S. Congressional Research Service Report for Congress 7-5700/R40778, Washington, DC. www.fas.org/sgp/crs/row/R40778.pdf.

Beintema, N., and G. J. Stads. 2010. *Public Agricultural R&D Investments and Capacities in Developing Countries—Recent Evidence for 2000 and Beyond*. IFPRI ASTI Background Note. Washington, DC: International Food Policy Research Institute. http://www.asti.cgiar.org/pdf/GCARD-BackgroundNote.pdf.

Brac de la PerriFre, R. A., and F. Seuret. 2000. *Brave New Seeds: The Threat of GM Crops to Farmers.* London: Zed Books.

Brambila-Macias, J., and I. Massa. 2010. "The Global Financial Crisis and Sub-Saharan Africa: The Effects of Slowing Private Capital Inflows on Growth." *African Development Review* 22: 366–377. doi: 10.1111/j.1467-8268.2010.00251.x.

Conner, A. J., T. R. Glare, and J.-P. Nap. 2003. "The Release of Genetically Modified Crops into the Environment." *Plant Journal* 33: 19–46. doi: 10.1046/j.0960-7412.2002.001607.x.

Easterly, W. 2009. "Can the West Save Africa?" *Journal of Economic Literature* 47 (2): 373–447.

Falck-Zepeda, J. 2009. "Socio-Economic Considerations, Article 26.1 of the Cartagena Protocol on Biosafety: What Are the Issues and What Is at Stake?" *AgBioForum* 12 (1): 90–107.

Falck-Zepeda, J., G. Traxler, and R. G. Nelson. 2000. "Surplus Distribution from the Introduction of a Biotechnology Innovation." *American Journal of Agricultural Economics* 82: 360–369.

FAO (Food and Agriculture Organization of the United Nations). 2004. *The State of Food and Agriculture. Agricultural Biotechnology: Meeting the Needs of the Poor?* Rome.

Finger, R., N. El Benni, T. Kaphengst, C. Evans, S. Herbert, S. Lehmann, S. Morse, and N. Stupak. 2011. "A Meta Analysis on Farm-Level Costs and Benefits of GM Crops." *Sustainability* 3: 743–762. doi: 10.3390/su3050743.

Gouse, M., J. F. Kirsten, B. Shankar, and C. Thirtle. 2005. "Bt Cotton in KwaZulu Natal: Technological Triumph but Institutional Failure." *AgBiotechNet* 7 (134): 1–7.

GRAIN. 2004. "Bt Cotton at Mali's Doorstep: Time to Act!" Accessed December 2010. www.grain.org/briefings_files/btcotton-synthesis-feb-2004-en.pdf.

Gruère, G. P., and D. Sengupta. 2009. "The Effects of GM-Free Private Standards on Biosafety Policymaking in Developing Countries." *Food Policy* 34 (5): 399–406.

Gruère, G. P., and H. Takeshima. 2012. "Will They Stay or Will They Go? The Political Influence of GM-Averse Importing Companies on Biosafety Decision Makers in Africa." *American Journal of Agricultural Economics* 94 (3): 736–749.

IFPRI (International Food Policy Research Institute). 2013. *Biotech and Biosafety Policy.* Accessed January 31, 2013. www.ifpri.org/ourwork/about/biotech-biosafety.

IMF (International Monetary Fund). 2009. *Impact of the Global Financial Crisis on Sub-Saharan Africa.* Washington, DC. www.imf.org/external/pubs/ft/books/2009/afrglobfin/ssaglobalfin.pdf.

James, C. 2011. *Brief 43: Global Status of Commercialized Biotech/GM Crops: 2011.* Ithaca, NY, US: International Service for the Acquisition of Agri-Biotech Applications.

Juma, C., and I. Serageldin. 2007. "Freedom to Innovate: Biotechnology in Africa's Development. A Report of the High-Level African Panel on Modern Biotechnology." Addis Ababa, Ethiopia, and Pretoria, South Africa: African Union and New Economic Partnership for Africa's Development. http://mrcglobal.org/files/Singh_Africa_20Year.pdf.

Karembu, M., F. Nguthi, and H. Ismail. 2009. "Biotech Crops in Africa: The Final Frontier." Nairobi, Kenya: ISAAA AfriCenter. www.isaaa.org/resources/publications/biotech_crops_in_africa/download/Biotech_Crops_in_Africa-The_Final_Frontier.pdf.

Knezevic, I. 2007. "Monsanto Rules: Science, Government, and Seed Monopoly." *Politics and Culture* 2. Accessed January 31, 2010. www.politicsandculture.org/2010/10/27/monsanto-rules-science-government-and-seed-monopoly/.

Moola, S., and V. Munnik. 2007. "GMOs in Africa: Food and Agriculture: Status Report 2007." Pretoria, South Africa: African Center for Biosafety. www.biosafetyafrica.org.za/images/stories/dmdocuments/gmos_in_africa.pdf.

Moschini, G., and H. Lapan. 1997. "Intellectual Property Rights and the Welfare Effects of Agricultural R&D." *American Journal of Agricultural Economics* 79 (4): 1229–1242.

Mulenga, D., and D. Shumba-Mnyulwa. 2010. "Overview of National Biosafety Frameworks with an Emphasis on Biosafety Socio-Economic Provisions." Presented at the Biosafety Socio-Economic Risk Assessment Training Workshop, February 15–19, in Pretoria, South Africa.

Munro, A. 2003. "Monopolization and the Regulation of Genetically Modified Crops: An Economic Model." *Environment and Development Economics* 8: 167–186. doi: 10.1017/S1355770X03000093.

Novy, A., S. Ledermann, C. Pray, and L. Nagarajan. 2011. "Balancing Agricultural Development Resources: Are GM and Organic Agriculture in Opposition in Africa?" *AgBioForum* 14 (3): 142–157.

Paarlberg, R. 2008. *Starved for Science: How Biotechnology Is Being Kept out of Africa.* Cambridge, MA, US: Harvard University Press.

Pew Initiative on Food and Biotechnology. 2004. *Feeding the World: A Look at Biotechnology and World Hunger.* Washington, DC.

Pontifical Academy of Sciences. 2010. "Science and the Future of Mankind." Accessed December 2010. www.vatican.va/roman_curia/pontifical_academies/acdscien/documents/sv%2099%285of5%29.pdf.

Potrykus, I., and K. Ammann, eds. 2010. "Proceedings of a Study Week of the Pontifical Academy of Sciences." Special issue. *New Biotechnology* 27 (5).

Pschorn-Strauss, E. 2005. "Bt Cotton in South Africa: The Case of the Makhathini Farmers." Accessed December 2010. www.grain.org/seedling_files/seed-05-04.pdf.

Qaim, M. 2009. "The Economics of Genetically Modified Crops." *Annual Review of Resource Economics* 1: 665–694.

———. 2010. "Benefits of Genetically Modified Crops for the Poor: Household Income, Nutrition, and Health." *New Biotechnology* 27 (5): 552–557.

Smale, M., P. Zambrano, G. Gruère, J. Falck-Zepeda, I. Matuschke, D. Horna, L. Nagarajan, et al. 2009. *Impacts of Transgenic Crops in Developing Countries during the First Decade: Approaches, Findings, and Future Directions*. IFPRI Food Policy Review 10. Washington, DC: International Food Policy Research Institute.

Spielman, D. J., and R. Pandya-Lorch, eds. 2009. *Millions Fed: Proven Successes in Agricultural Development*. Washington, DC: International Food Policy Research Institute.

Stone, G. D. 2002. "Both Sides Now: Fallacies in the Genetic-Modification Wars, Implications for Developing Countries, and Anthropological Perspectives." *Current Anthropology* 43 (4): 611–619.

———. 2011. "Field vs. Farm in Warangal: Bt Cotton, Higher Yields, and Larger Questions." *World Development* 39 (3): 387–398.

Takeshima, H., and G. Gruère. 2011. "Pressure Group Competition and GMO Regulations in Sub-Saharan Africa—Insights from the Becker Model." *Journal of Agricultural and Food Industrial Organization* 9 (7).

Tripp, R. 2009. *Biotechnology and Agricultural Development: Transgenic Cotton, Rural Institutions and Resource-Poor Farmers*. Oxon, UK: Routledge.

Uzogara, S. G. 2000. "The Impact of Genetic Modification of Human Foods in the 21st Century: A Review." *Technology Advances* 18 (3): 179–206.

World Bank. 2007. *World Development Report 2008: Agriculture for Development*. Washington, DC: International Bank for Reconstruction and Development/World Bank.

Chapter 1

Socioeconomic and Farm-Level Effects of Genetically Modified Crops: The Case of Bt Crops in South Africa

Marnus Gouse

The year 2011 was the 14th since the first commercial release of a genetically modified (GM) crop in South Africa. In 1997/98, insect-resistant (Bt) cotton was released for production, and South Africa became the first country in Africa where a GM crop was produced on a commercial level. Bt maize was approved for commercial production in 1998/99, and Bt yellow maize was planted in the same season. The first plantings of Bt white maize in 2001/02 established South Africa as the first GM subsistence crop producer in the world. Herbicide-tolerant (HT) cotton was made available for commercial production in the 2001/02 season along with HT soybeans. Commercialization of HT maize seeds followed in 2003/04. GM cotton containing the combined or "stacked" trait (Bt and HT) was released for the 2005/06 season, and Bt/HT maize was released for the 2007/08 production season.

This chapter supplies a brief summary of the performance, socioeconomic impacts, and main issues surrounding Bt cotton and GM maize in South Africa. A substantial number of peer reviewed papers on GM crops in South Africa have been published, and it is recommended that interested readers refer to these publications for more in-depth information and discussion on the studies and findings.

South African Biosafety Framework

In 1989 a US seed company approached the South African Department of Agriculture for permission to perform contained field trials with Bt cotton. This set in motion the South African biosafety regulatory process and initiated the first trials with GM crops on the African continent. The South African Committee for Genetic Experimentation (SAGENE) had been formed in 1979 by public and private scientists to monitor and advise the National Department of Agriculture and industry on the responsible development of genetically modified organisms (GMOs) through the provision of guidelines and the approval of research centers and projects. SAGENE gained statutory status in 1992 as the national advisory committee on modern GM biotechnology. The approval for the commercial release of Bt cotton and maize was done under the guidelines of SAGENE for the 1997/98 and 1998/99 seasons. These guidelines and procedures remained the biosafety framework cornerstone until South Africa's GMO Act 15 of 1997 was approved by Parliament in June 1997 and entered into force in November 1999, when the regulations were published. In 1999 SAGENE was replaced by the scientific Advisory Committee that was established under the GMO Act (Wolson and Gouse 2005). The South African GMO Act 15/1997, as amended in 2006, provides a comprehensive biosafety framework to manage research, development, application, production, and trade in GMOs. The GMO Secretariat is housed in the Department of Agriculture, and decisionmaking is vested in the GMO Executive Council that represents eight government departments. The Council is advised by a national Advisory Committee of scientific experts.

Since implementation, the GMO legislation has served the country well in its balanced approach to modern biotechnology and its applications. However, more recently there have been some unclear delays in the decisionmaking process, and the scientific community and academia have expressed concern that decisionmaking has become less scientific and a lack of transparency in the process could lead to an increase in the cost of regulation and in the opportunity cost for research institutions, innovators, and in reality, consumers.

Bt Cotton

In 2007 GM cotton globally covered 15 million hectares (43 percent of total world cotton), of which Bt varieties accounted for 10.8 million hectares and a further 3.2 million hectares as Bt combined with a second Bt or with an herbicide-tolerance trait (James 2007). In 2009 the global GM cotton area increased to 16.2 million hectares and in 2011 to 25 million or 68 percent of global cotton plantings (James 2009, 2011). Historically, cotton has been responsible for about 25 percent of global chemical insecticides used in agriculture due to attacks by a range of insect pests (Woodburn 1995), with cotton bollworm being the main pest. In an effort to reduce insecticide use and with insect resistance build-up against chemicals, Bt technology has offered a cost-saving and environmentally friendlier alternative.

Cotton planting in South Africa declined from its peak of 180,000 hectares in 1988 (under tariff protection) to just over 5,000 hectares in 2010 due to a combination of market liberalization, low world cotton prices, and relatively better prices for competing crops like maize, sunflower seed, and sugar cane. South Africa has been a net importer of cotton for the past couple of decades. In 1997/98 South Africa became the first country in Africa to commercially produce GM crops with the release of Bt cotton. The initial uptake of the first Bt cotton varieties of the US cotton seed company, Delta and Pineland (D&PL), was less than spectacular, as the conventional varieties of local ginning companies were more popular. Some commercial farmers were also cautious during the first seasons and wanted to test the new technology and see how ginners and the rest of the industry reacted. However, when the Bt gene was introduced into D&PL's popular OPAL variety (originally from Australia), adoption increased dramatically. NuOPAL (Bt), DeltaOPAL RR (HT), and NuOPAL RR (Stacked Bt/HT), which are currently planted in South Africa, are all based on the Delta OPAL germplasm (Gouse 2009).

As clearly shown in Table 1.1, Bt cotton has been very popular, reaching 70 percent of total cotton area in 2003. The share decreased somewhat with the introduction of HT cotton, but Bt cotton remained the more popular of the two. With the introduction of stacked cotton (with both the Bt and HT events), Bt's share dropped considerably as farmers opted for cotton with both traits. By the 2005/06 season 92 percent of the cotton plantings in South Africa were GM. A large share of the conventional cotton being planted is mandatory refugia that are planted alongside Bt fields to prevent insect-resistance development. Farmers tend to plant HT cotton as refugia for stacked Bt/HT plantings.

Despite various land reform and development projects attempting to settle small-scale farmers in established and potential cotton production areas, the traditional areas of Tonga (in Kangwane Mpumalanga) and Makhathini Flats (KwaZulu-Natal) remain the major contributors to smallholder cotton production. The total number of smallholder cotton producers has varied but generally amounts to a few thousand farmers with the vast majority of them situated on the Makhathini Flats. As large-scale farmers produce the bulk of the South African cotton crop, it would not be totally correct to suggest that the adoption figures in Table 1.1 apply to smallholders as well, though Bt cotton adoption by smallholders has not been less impressive. In the first commercialization season of 1997, only 4 farmers planted demonstration Bt plots under the guidance of Monsanto (the technology owner). In 1998, 75 farmers, or 3.4 percent of the cotton farmers on Makhathini, planted Bt cotton; in 1999, 411 farmers, or 13.7 percent, planted Bt. In 2000, 1,184 cotton farmers (39.5 percent) on the Makhathini Flats planted Bt cotton. In 2001 it was

TABLE 1.1 Estimated area and share of total area planted to transgenic cotton in South Africa, 2000/2001–2007/08

Event	2000/2001	2001/02	2002/03	2003/04	2004/05	2005/06	2006/07	2007/08
Bt cotton (hectares)	12,470	14,700	15,800	20,700	12,719	7,060	2,500	780
Bt cotton (%)	22	38	70	58	60	39	22	6
Cotton (hectares)	0	3,868	3,386	9,280	6,360	2,350	113	520
HT cotton (%)	0	10	15	26	30	13	10	4
Stacked (Bt/HT) cotton (hectares)	0	0	0	0	0	7,240	6,820	10,660
Stacked cotton (Bt/HT) (%)	0	0	0	0	0	40	60	82
Total share planted to transgenic cotton (%)	22	48	85	84	90	92	92	92

Source: Gouse, Kirsten, and Van der Walt (2008).

Notes: Figures in previous publications have been updated with additional and revised figures. Official figures for the most recent years were not available at time of writing. Unofficial indications suggest that in 2012 nearly 100% of cotton produced in South Africa was GM, with stacked cotton making up the major share and farmers planting HT cotton as mandatory refugia. Bt = insect resistant; HT = herbicide tolerant.

estimated that close to 3,000 of the 3,229 farmers on the Flats planted Bt, reaching close to 90 percent adoption in five years (Gouse 2009).

This remarkable adoption rate was explained partly by the impressive performance of Bt cotton as planted by the first adopting Makhathini farmers. However, the other major explanation was that the sole credit and input supplier and cotton buyer on the Flats, Vunisa, also noticed the performance of Bt cotton and started to recommend the seed to its clients/farmers. As the main objective of a cotton gin is to gin as much cotton as possible, Vunisa wanted to increase the cotton crop on the Flats but not at the expense of their credit book. After monitoring the performance of Bt cotton for the first couple of seasons, Vunisa decided that it could increase the ginable cotton crop, and decrease the risk of crop failure (due to bollworm damage) and thus their credit risk by recommending Bt cotton to farmers. It can be argued that even though Vunisa was making inputs available to farmers under credit long before Bt was introduced, the availability of credit and the role Vunisa's extension officers had in recommending Bt seed played a large role in smallholders' ability and decision to adopt the new technology (Gouse 2009).

All the peer reviewed publications on Bt cotton in South Africa (mainly focusing on smallholder farmers) report yield increases with the use of Bt cotton compared to conventional varieties (Table 1.2). Almost all studies also showed savings in insecticide expenditure; with the exception of results from the one-year, 20-farmer study by Hofs, Fok, and Vaissayre (2006). Even though most of the yield differences were substantial, some were found not to be statistically significant, mainly due to small sample sizes and large variability in the data. Compared to study results in countries like Australia, China, India, and Mexico, the relative yield gain from the use of Bt cotton in South Africa is higher. One of the reasons for this is that the base yield (non-Bt cotton) of smallholders is very low, and a small change in yield is exaggerated when expressed relative to a low conventional variety yield. In fact, in some other countries, the yield advantage of Bt cotton was more than the total seed cotton yield attained per hectare in South Africa (Fok et al. 2007). Gouse, Kirsten, and Jenkins (2003) found an 18.5 percent yield increase for South African large-scale irrigation farmers for the 2000/2001 season, which compares well with a 16.8 percent increase measured on field trials at a Clark Cotton (a ginning company) experimental farm in Mpumalanga. Large-scale dryland farmers enjoyed a 14 percent yield increase, while some studies found that small-scale dryland farmers enjoyed an increase of between 23 and 85 percent over a number of seasons (Table 1.2).

TABLE 1.2 Summary of findings of main published studies

Type of farm and year	Yield (MT/ha)			Cost of seed (US$/ha)			Cost of insecticide (US$/ha)			Difference in gross margin (US$/ha)
	Non Bt	Bt	Difference (%)	Non Bt	Bt	Difference (US$/ha)	Non Bt	Bt	Difference (US$/ha)	
Smallholder										
1999/2000[a]	395	576	46	n.a.	n.a.	−26	20	15	5	58
1998/99[b]	452	738	63	23	46	−23	25	12	13	88
1999/2000	264	489	85	30	65	−35	35	16	19	61
2000/2001	501	783	56	23	34	−11	40	15	25	96
2002/03[c]	423	522	23	23	44	−21	32	23	9	23
Large-scale										
Dryland 2000/2001[a]	832	947	14	n.a.	n.a.	−30	25	10	15	25
Irrigation 2000/2001	3,413	4,046	19	n.a.	n.a.	−54	67	29	38	209

Source: Adjusted from Gouse (2009).

Notes: US$ calculated using average South African rand/US$ exchange rate for September–May for applicable years. MT = metric tons; n.a. = not available; US$ = US dollars.

[a]Gouse, Kirsten, and Jenkins (2003).

[b]Bennett, Morse, and Ismael (2006).

[c]Fok et al. (2007).

These trends are consistent with findings elsewhere, such as in Argentina (Qaim, Cap, and De Janvry 2003), where large-scale commercial farmers were reported to enjoy 19 percent yield increases and smallholder farmers reported 41 percent yield increases. Like Qaim, Cap, and De Janvry (2003), South African researchers attribute the difference between the Bt yield advantages of small- and large-scale farmers to the financial and human capital constraints that cause smallholders to invest in chemical pest control. Shankar and Thirtle (2005) showed that the average insecticide application level of smallholder farmers on the Makhathini Flats is lower than 50 percent of the optimal level; it is thus not surprising that Bt cotton is able to substantially reduce the yield loss caused by bollworms. With low control-group yields and limited (and in many cases in-effective) chemical insecticide applications, exaggerated yield increases in excess of 50, 60, and 80 percent as reported by Bennett, Morse, and Ismael (2006) do not seem so mind-boggling. But these results have to be seen in context, and as the authors caution, the figures might also be inflated due to selection bias.

The yield increase with Bt cotton, compared to conventional cotton, depends on the bollworm infestation level in the particular season and the effectiveness of chemical bollworm control by the farmer. It can be expected that the yield advantage will differ across farmers, farms, regions, and seasons (Fok et al. 2007). Both large-scale and smallholder farmers enjoyed significant savings on insecticides (generally 3/4/5 pyrethroid sprays), and despite higher expenditure on seed (as a result of the additional technology fee), they enjoyed a higher gross margin. However, it is important to stress that Bt does not kill all insects, and chemical spraying is still required to prevent damage by sucking insects, which in the past have been killed in the cross-fire aimed at bollworms.

The Bt technology fee was adjusted downward by about 24 percent after the introduction season, following farmer concerns that the technol-ogy was not affordable. The fee was then held constant at South African rand (ZAR) 600 per 25 kilograms of seed (between about $50 and $75 according to the fluctuating local currency)[1] for 1999/2000–2002/03, at ZAR700 for 2003/04–2004/05, and then at ZAR785 from 2005/06 to the 2008/09 season. Between 1999 and 2008 a 25 kilogram bag of conventional cotton sold for between ZAR150 and ZAR430. This means that the extra Bt technology fee per 25 kilogram bag was between 1.8 and 4.0 times the price of the bag of seeds (Gouse 2009).

Analysis of "who gains?" from Bt technology showed that despite the high technology fee, farmers captured the lion's share of the additional benefits

---

1   All dollar amounts are US dollars.

generated by the introduction of this new technology (Gouse, Pray, and Schimmelpfennig 2004). Basing their calculations on the abovementioned studies, Brookes and Barfoot (2010) estimated that in the 11 years from 1998 to 2008, the use of Bt cotton contributed an additional $21 million to farm income in South Africa.

The Makhathini Flats smallholder experience with Bt cotton has been hailed internationally as the first example of how modern biotechnology can benefit resource-poor farmers in Africa. There can be no doubt that the majority of Makhathini Flats farmers did indeed benefit from the introduction of Bt cotton. They were able to adopt and benefit from this new technology because all the institutional structures that facilitate a functioning market were in place at the time. These structures include functioning input markets (credit, seeds, and chemicals) and output markets (seed cotton buyer) that operate at market clearing prices. An important factor was that Vunisa was the only buyer and, because of this monopsony power, could supply production credit to farmers who did not own their land, using the forthcoming crop as collateral (Gouse, Shankar, and Thirtle 2008). This system is not uncommon to Africa, where widespread failure of credit and input markets (partly due to lack of land ownership that could serve as collateral) has led to interlocked transactions, in which a firm wishing to purchase the farm output—typically a ginner in the case of cotton—provides inputs to farmers on credit and attempts to recover the credit upon purchase of the product (Tschirley, Poulton, and Boughton 2006). However, when the credit system collapsed in 2002—because of farmers defaulting on their loans as a consequence of a combination of droughts, low prices (linked to the low and stagnated world cotton price), marginal profits, adverse selection, and market competition—the whole system collapsed, and cotton production dropped.

The Makhathini smallholder experience is indeed a good example for the rest of Africa, as countries considering adoption of Bt cotton need to take note that although technical solutions can help address problems (such as lack of knowledge regarding insects and pest control, limited access to inputs, or evolution in pest pressure), no technology (GM or otherwise) can resolve the fundamental institutional challenges of smallholders and agriculture in Africa. The particular case of the Makhathini Flats and the wider story of cotton in South Africa emphasize that although all agricultural systems require adequate investment and appropriate technologies, their viability is determined by the policies and institutions that facilitate sustainable and profitable production. Bt cotton and more recently stacked (Bt/HT) varieties are still the varieties of choice for smallholder producers, but production levels have decreased drastically and remain limited mainly due to the relatively low price of cotton.

Bt Maize

Globally, in 2007 GM maize was planted on 35 million hectares, or 24 percent of world maize plantings, of which 9.3 million hectares was Bt as single trait and another 18.8 million hectares in combination with other traits (James 2007). In 2010 GM maize covered 46.8 million hectares globally, and the area increased to 51 million hectares in 2011 (James 2010, 2011). Bt maize was first introduced in the United States in 1996, and by 2006 it covered 40 percent or 12.7 million hectares of the total US maize crop. In Argentina, varieties containing the Bt trait were planted on 73 percent of the total Argentinean maize area, and in Spain it covered 54,000 hectares or 15 percent of the total maize area (Brookes and Barfoot 2008).

Maize is the most important field crop in South Africa and annually covers an estimated 30 percent of the total arable land. Maize serves as staple food for the majority of the South African population and also as the main feedgrain for livestock. Between 60 and 70 percent of the South African yellow maize production is consumed in the chicken-production sector. Over the past 9–10 years, South Africa produced an average of 9.3 million metric tons of maize on 2.75 million hectares.

Even though Bt yellow maize was released in 1998 for commercial production, GM white maize was commercialized only in 2001. That year, South Africa became the first country in the world to permit the commercial production of a GM subsistence crop—Bt white maize. In South Africa and other southern African countries, the losses sustained in maize crops due to damage caused by the African maize stem (stalk) borer (*Busseola fusca*) are estimated to be between 5 and 75 percent, and it is generally accepted that, pre-Bt, *Busseola* annually reduced the South African maize crop by an average of 10 percent (Annecke and Moran 1982). Gouse et al. (2005) showed that in 2005 with a seemingly conservative estimate of 10 percent for damage caused by both *Busseola fusca* and *Chilo partellus,* the average annual loss (in the absence of Bt) adds up to just under a million tons of maize, with an approximate value of ZAR810 million. At the 2008 maize price level (more or less similar to the 2011 price level), the potential damage caused by borers would be closer to ZAR1.6 billion (about $200 million). Both *B. fusca* and *C. partellus* can be controlled to a satisfactory level with the use of the *Bt* gene currently used in South African Bt varieties (Cry1Ac).

As can be seen in Table 1.3, the initial spread of Bt maize was quite slow because of the scale-up time required to have a sufficient amount of seeds and to have the Bt trait inserted in hybrids that were suitably adapted to local conditions. Approval for commercial release of herbicide tolerance came in 2002 and the stacked traits of Bt and HT in 2007. Compared to cotton, the decrease in Bt and HT maize since the introduction of stacked maize was less pronounced. Bt remains the most popular trait, partly because especially white stacked maize adoption has been hindered by inadequate seed availability. In the 2008/09 production season, GM maize covered 70 percent of the total South African maize area, with Bt maize covering 43 percent. In 2009/10 the Bt maize area increased by a further 269,000 hectares up to 48 percent, mainly stemming from a drop in the white stacked maize area because of inadequate seed supply.

Considering the adoption rates illustrated in Table 1.3, it is possible to con-clude that South African maize farmers have benefited from the introduction of GM maize. Similar to the indicated GM cotton adoption rates in Table 1.1, these GM maize adoption rates represent adoption by predominantly com-mercial farmers. There are no official smallholder GM maize adoption figures, but it is estimated that about 10,500 subsistence, smallholder, and emergent farmers (about 23 percent of the smaller farmers), buying hybrid seed from the three major seed companies, planted GM maize in 2007 (Gouse, Kirsten, and Van der Walt 2008). However, there are still areas in South Africa where small-holders plant mainly open-pollinated varieties and traditional/saved seed, and definitions of subsistence, smallholder, smallholder projects, and emerging farmers also complicate estimations. It can therefore be argued that the num-ber of smallholders planting GM maize is still relatively minimal.

Marra, Pardey, and Alston (2002) found that there were significant ben-efits to planting Bt maize in the United States through increased yields, even when it appeared as if borer infestation levels were not large enough to con-trol with insecticides. Marra, Carlson, and Hubbell (1998) reported that the use of Bt maize boosted yields by 4–8 percent, depending on location and year. Results from outside the United States show a similar pattern. In the Huesca region in Spain, Brookes (2002) reported a yield increase of 10 percent over conventional maize protected with pesticides and an increase of 15 per-cent when insecticides were not used. Other regions in Spain enjoyed an aver-age Bt yield advantage of 6.3 percent, with a range of 2.9–12.9 percent. James (2002) reported a 8–10 percent yield increase in Argentina up to 2004, and more recent studies show a 5–6 percent increase (Brookes and Barfoot 2008). Gonzales (2002) recorded a yield advantage of 41 percent for Bt maize on

TABLE 1.3   Estimated area and share of total area planted to genetically modified maize in South Africa, 2000/2001–2009/10

Event	2000/2001	2001/02	2002/03	2003/04	2004/05	2005/06	2006/07	2007/08	2008/09	2009/10
Bt yellow maize (hectares)	59,000	160,000	176,000	197,000	249,000	107,000	391,000	406,000	376,000	267,000
Bt yellow maize (%)	5.0	14.5	19.5	19.7	22.6	17.8	35.5	38.2	4C.1	26.1
Bt white maize (hectares)	0	6,000	60,000	144,000	142,000	221,000	712,000	696,000	660,00C	1,038 000
Bt white maize (%)	0	0.4	2.8	8.0	7.9	22.8	43.8	40.1	44.3	60.3
HT yellow maize (hectares)	0	0	0	0	14,000	68,000	137,000	159,000	159,000	-54,000
HT yellow maize (%)	0	0	0	0	1.3	11.3	12.5	15.0	16.9	15.1
HT white maize (hectares)	0	0	0	0	5,000	60,000	139,000	218,000	160,000	91,000
HT white maize (%)	0	0	0	0	0.3	6.0	8.5	13.0	10 7	5.3
Stacked (Bt/HT) yellow maize (hectares)	0	0	0	0	0	0	0	23,000	107,000	201,000
Stacked (Bt/HT) yellow maize (%)								2.0	11 4	19.6
Stacked (Bt/HT) white maize (hectares)								60,000	226,000	139,000
Stacked (Bt/HT) white maize (%)								3.0	15.2	8.1

Source: Updated from Gouse, Kirsten, and Van der Walt (2008).
Note: Bt = insect resistant; HT = herbicide tolerant.

field trials in the Philippines, and Philippine farmers indicated an even higher (60 percent) yield improvement. In most countries, the additional cost of the Bt technology has exceeded the savings on insecticides and thus has resulted in an increase in total production costs.

Compared to the number of studies and publications on Bt cotton in South Africa, the body of literature and the number of researchers following the farm-level impacts of Bt maize in South Africa is rather limited. Even though there have been reports in the media quoting some anecdotal findings of some fairly unscientific studies, only a series of studies by the University of Pretoria[2] have endeavored to follow the socioeconomic effects and performance of Bt maize for a number of seasons, mainly focusing on smallholders.

Gouse et al. (2005) found average yield increases (due to better stem borer control) of 10–11 percent for commercial (dryland and irrigation) farmers, whereas smallholder Bt adopters reported yield increases of 0–32 percent for the seven seasons 2001/02–2007/08 (Gouse et al. 2010). A statistically insignificant average yield increase of 12 percent was found across the seven seasons. In seasons with a low stem borer infestation, resulting in insignificant stalk borer damage, farmers planting Bt maize seed were in all likelihood worse off than farmers planting conventional hybrid maize because of the extra Bt technology fee. It is however difficult to make preplanting predictions on seasonal stalk borer infestation levels due to the complicated relationship between rainfall, variable seasons, growth of maize, effect of stalk borer on the maize plant, and the effect of natural enemies on the host (Annecke and Moran 1982). Because a dry early season does not necessarily portend a dry season throughout, South African large-scale commercial farmers indicate that Bt serves as affordable insurance against unforeseeable stalk borer outbreaks, but increases in seed cost or technology fees could easily outstrip that insurance value to small-scale and subsistence farmers in South Africa (Gouse et al. 2006).

Gouse et al. (2006) endeavored to quantify the 16 percent yield increase, the average of the Bt yield advantage for two groups of farmers in northern KwaZulu-Natal for the 2002/03 season, in subsistence-farmer terms. For these smallholders a 16 percent yield increase meant only 110 kilograms of extra grain, and selling the extra grain would render a rather insignificant income advantage. However, arguing that the extra grain replaces potentially purchased, relatively more expensive, maize meal (flour), the yield

---

2 Mainly supported by the Rockefeller Foundation and the Economic and Social Research Council/ Department for International Development funding and in collaboration with, among others, Rutgers University, Imperial College, and the Programme on Mycotoxins and Experimental Carcinogenesis at the South African Medical Research Council.

advantage seems more valuable. Alternatively, using a generally excepted rule-of-thumb stating that a rural household of seven members requires fourteen 80 kilogram bags of maize meal for a year to be food secure, the 16 percent yield advantage in 2002/03 resulted in approximately 36 more days of maize meal for the household. This is assuming, rather unrealistically, that there is no postharvest damage to harvested grain.

Insecticide use by maize-producing smallholders is limited, and Bt adoption consequently did not result in substantial insecticide savings. Bt adoption by commercial farmers has resulted in decreased expenditure on insecticides but, similar to what has been reported in other countries, generally not enough to cover the increased seed cost. Depending on the quantity of seed purchased, Bt maize seed was 23–25 percent more expensive than conventional seed, and more recently those percentages have increased to about 27–30 percent.

Following the planting of HT maize demonstration plots in 2003/04 and 2004/05 in some smallholder areas where Bt had been introduced, a number of farmers adopted HT maize in 2005/06. Many farmers who planted Bt maize in previous years instead opted for HT seed (Gouse et al. 2010). Farmers indicated that compared to stem borers, weeds are a constant pest, and it would seem as if the labor-saving benefit of HT maize is valued higher than the insect control (yield) benefit of Bt. With a substantial share of the economically active, able-bodied population emigrating to urban areas in search of employment and a tragically high HIV/AIDS prevalence, especially in rural KwaZulu-Natal, labor has become a scarce commodity for many South African smallholder farmers. By using broad-spectrum herbicides before and after planting (some only after), as opposed to manual weeding with hand and hoe, farmers are able to save quite considerably on family labor person-days. For some of the seasons, HT maize also yielded more grain than conventional hybrid maize with manual weeding because of more effective weed control. In some areas in KwaZulu-Natal, HT has totally replaced Bt. In an attempt to benefit from both the GM technologies, some smallholders purchased stacked Bt/HT maize, but others indicated this maize is too expensive and opted for HT only (Gouse et al. 2010). Stem borer pressure has been low during the study seasons, and it would be interesting to see how HT-adopting farmers react to possible higher borer levels in seasons to come.

Using yield increase, insecticide savings, and increased seed expenditure indications of mainly Gouse et al. (2005), Brookes and Barfoot (2010) estimated that between 2000 and 2008, Bt maize adoption increased adopting farmers' farm income by a total of $476 million. That is a productivity increase–induced injection into the economy of ZAR3.67 billion over 9 years. To put this amount into context, over the past 10 years the South African Government has, through the platforms and initiatives created under the National Biotechnology Strategy (Republic of South Africa, Department of Science and Technology 2001), invested about ZAR900 million (Hanekom 2010) in biotechnology research and development.

Conclusion

Benefiting from a strong research background, South Africa was able to proactively develop guidelines and later legislation and regulations on the development and use of modern biotechnology and its applications such as GM crops. Development and implementation of a relatively dynamic GMO legislation and underlying regulations have enabled South African farmers—and, to a lesser extent, consumers, through maize meal prices and health aspects (see Chapter 2)—to benefit from the first wave of GM crops.

Solely based on the high adoption levels of especially Bt cotton and maize by large-scale farmers, in the presence of available and less-expensive conventional seed varieties (including near isolines), it is possible to conclude that farmers benefited. Some peer-reviewed studies have shown that like large-scale farmers, smallholder cotton and maize farmers have also benefited, mainly through savings on insecticide applications and limitation of the damage caused by bollworms and stem borers.

Whereas Bt cotton saw a near 100 percent smallholder adoption rate in only a couple of years, adoption of Bt maize has been limited. There are a number of reasons for this: in a vertically integrated production system where the input supplier also ensures an output market, adoption of a (early season) more expensive but productivity-increasing technology makes sense. However, smallholder maize farmers have to fund production inputs, and as many only produce on a subsistence level (in many cases surplus production depends on the season's rainfall), farmers are unable to directly recover their input expenditures. Contrary to cotton, for which bollworm pressure and damage seems to be more constant and severe, stem borer infestation levels (especially on dryland maize) vary significantly from season to season and across areas, and the damage level is generally lower than with cotton. Though very few smallholder maize farmers apply an insecticide to control stem borers on maize, the amount of labor and chemicals required to control borers is far less than what is required to control bollworms on cotton. Another factor that is sometimes not taken into consideration, especially in the South African context, is that smallholder maize farmers' reasons or motivation for maize production differ. In the production of a cash crop like cotton, farmers are profit driven and are intent on producing as much as possible. On the contrary, some smallholder maize farmers are only interested in producing enough for their households, others only plant a couple of lines for fresh maize, and yet others only produce to sell. It is unlikely that the smaller producers would invest in a productivity-increasing technology like Bt maize.

That a technology was introduced and adopted and that farmers benefited does not necessarily result in a flourishing sector, as is evident from the example of South African cotton. Even with biotechnology, South African cotton farmers were not able to produce profitably at low cotton world prices. The fact that many smallholders continued producing, while commercial farmers left the sector for greener or more profitable pastures is indicative of smallholders' dependence on government support and limited alternative production options and not of the success of biotechnology.

Bt seed technology is a production tool just like fertilizers, herbicides, or irrigation technologies. Contrary to the technologies of the Green Revolution, it might be able to improve the yields of farmers with limited ability or means to control insects. However, it will by no means be able to overcome institutional failure and governance challenges that seem to be endemic in African agriculture and that were also the limiting factors in the Green Revolution. The experience with Bt cotton on the Makhathini Flats emphasizes that technology-induced advances might be short lived in the absence of the correct institutional structures, regulations, cooperation, and competition.

References

Annecke, D. P., and V. C. Moran. 1982. *Insects and Mites of Cultivated Plants in South Africa*. Durban, South Africa: Butterworths.

Bennett, R., S. Morse, and Y. Ismael. 2006. "The Economic Impact of Genetically Modified Cotton on South African Smallholders: Yield, Profit and Health Effects." *Journal of Development Studies* 42: 662–677.

Brookes, G. 2002. *The Farm-Level Impact of Using Bt Maize in Spain*. Brussels: EuropaBio.

Brookes, G., and P. Barfoot. 2008. *GM Crops: Global Socio-Economic and Environmental Impacts 1996–2006*. Dorchester, UK: PG Economics.

———. 2010. *GM Crops: Global Socio-Economic and Environmental Impacts 1996–2008*. Dorchester, UK: PG Economics.

Fok, M., J. L. Hofs, M. Gouse, and J. F. Kirsten. 2007. "Contextual Appraisal of GM Cotton Diffusion in South Africa." *Life Sciences International Journal* 1 (4): 468–482.

Gonzales, L.A. 2002. "Likely Transcendental Effects of Agribiotechnology: The Case of Bt Hybrid Corn in the Philippines." Paper presented during the Symposium on Bt Technology: Facts and Issues, May 5, Los Baños, Laguna, Philippines.

Gouse, M. 2009. "Ten Years of Bt Cotton in South Africa: Putting the Smallholder Experience into Context." In *Biotechnology and Agricultural Development: Transgenic Cotton, Rural Institutions and Resource-Poor Farmers,* edited by R. Tripp, 200–224. New York: Routledge.

Gouse, M., J. F. Kirsten, and L. Jenkins. 2003. "Bt Cotton in South Africa: Adoption and the Impact on Farm Incomes amongst Small-Scale and Large-Scale Farmers." *Agrekon* 42 (1): 15–28.

Gouse, M., J. F. Kirsten, and W. J. Van Der Walt. 2008. "Bt Cotton and Bt Maize: An Evaluation of Direct and Indirect Impact on the Cotton and Maize Farming Sectors in South Africa." Commissioned report to the Department of Agriculture: Directorate BioSafety, Pretoria, South Africa.

Gouse, M., J. Piesse, and C. Thirtle. 2006. "Output and Labour Effects of GM Maize and Minimum Tillage in a Communal Area of KwaZulu Natal." *Journal of Development Perspectives* 2 (2): 192–207.

Gouse, M., C. E. Pray, and D. E. Schimmelpfennig. 2004. "The Distribution of Benefits from Bt Cotton Adoption in South Africa." *AgBioForum* 7 (4): 187–194.

Gouse, M., B. Shankar, and C. Thirtle. 2008. "The Decline of Cotton in KwaZulu Natal: Technology and Institutions." In *Hanging by a Thread: Cotton, Globalization, and Poverty in Africa*, edited by W. G. Moseley and L. C. Gray, 103–120. Athens, OH, US: Ohio University Press.

Gouse, M., C. E. Pray, J. F. Kirsten, and D. E. Schimmelpfennig. 2005. "A GM Subsistence Crop in Africa: The Case of Bt White Maize in South Africa." *International Journal of Biotechnology* 7 (1–3): 84–94.

Gouse, M., C. E. Pray, D. E. Schimmelpfennig, and J. F. Kirsten. 2006. "Three Seasons of Subsistence Insect-Resistant Maize in South Africa: Have Smallholders Benefited?" *AgBioForum* 9 (1): 15–22.

Gouse, M., J. F. Kirsten, J. Piesse, C. Thirtle, and C. Poulton. 2010. "Insect Resistant and Herbicide Tolerant Maize Adoption by South African Smallholder Farmers—Making Sense of Seven Years of Research." Presented at the 14th International Consortium on Applied Bioeconomy Research conference, BioEconomy Governance: Policy, Environmental and Health Regulation and Public Investments in Research, June 16–18, in Ravello, Italy.

Hanekom, D. A. 2010. South African Deputy Minister of Science and Technology at launch of Biosafety South Africa, February 18, in Somerset Wes, South Africa. www.bizcommunity.com/Article/196/148/45017.html.

Hofs, J. L., M. Fok, and M. Vaissayre. 2006. "Impact of Bt Cotton Adoption on Pesticide Use by Smallholders: A 2-Year Survey in Makhathini Flats (South Africa)." *Crop Protection* 25: 944–988.

James, C. 2002. *Global Review of Commercialized Transgenic Crops 2001: Feature Bt Cotton.* Ithaca, NY, US: International Service for the Acquisition of Agric-Biotech Applications.

———. 2007. *Global Status of Commercialized Biotech/GM Crops: Brief 37-2007.* Ithaca, NY, US: International Service for the Acquisition of Agric-Biotech Applications.

———. 2009. *Global Status of Commercialized Biotech/GM Crops: Brief 39-2009.* Ithaca, NY, US: International Service for the Acquisition of Agric-Biotech Applications.

———. 2010. *Global Status of Commercialized Biotech/GM Crops: Brief 42-2010.* Ithaca, NY, US: International Service for the Acquisition of Agric-Biotech Applications.

———. 2011. *Global Status of Commercialized Biotech/GM Crops: Brief 43-2011.* Ithaca, NY, US: International Service for the Acquisition of Agric-Biotech Applications.

Marra, M. C., G. Carlson, and B. Hubbell. 1998. *Economic Impacts of the First Crop Biotechnologies.* Raleigh, NC, US: North Carolina State University.

Marra, M. C., P. G. Pardey, and J. M. Alston. 2002. "The Payoffs to Agricultural Biotechnology: An Assessment of the Evidence." Environment and Production Technology Division Discussion Paper 87. Washington, DC: International Food Policy Research Institute.

Qaim, M., E. J. Cap, and A. De Janvry. 2003. "Agronomics and Sustainability of Transgenic Cotton in Argentina." *AgBioForum* 6 (1–2): 41–47.

Republic of South Africa, Department of Science and Technology. 2001. *National Biotechnology Strategy for South Africa.* Pretoria. http://www.esastap.org.za/download/sa_biotechstrat_jun2001.pdf.

Shankar, B., and C. Thirtle. 2005. "Pesticide Productivity and Transgenic Cotton Technology: The South African Smallholder Case." *Journal of Agricultural Economics* 56 (1): 97–115.

Tschirley, D., C. Poulton, and D. Boughton. 2006. *The Many Paths of Cotton Sector Reform in Eastern and Southern Africa: Lessons from a Decade of Experience.* International Development Working Paper 88. East Lansing, MI, US: Michigan State University.

Wolson, R. A., and M. Gouse. 2005. "Towards a Regional Approach to Biotechnology Policy in Southern Africa: Phase I, Situation and Stakeholder Analysis—South Africa." Food, Agriculture and Natural Resources Policy Analysis Network [FANRPAN] draft paper. Pretoria, South Africa: FANRPAN.

Woodburn, A. 1995. *Cotton: The Crop and Its Agrochemicals Market.* Balerno, Scotland: Allen Woodburn Associates / Managing Resources.

Chapter 2

Bt Maize and Fumonisin Reduction in South Africa: Potential Health Impacts

Carl E. Pray, John P. Rheeder, Marnus Gouse, Yvette Volkwyn,
Liana van der Westhuizen, and Gordon S. Shephard

Starting in the late 1940s, deaths from esophageal cancer (EC) rose dramatically in rural areas of the Transkei region of South Africa, which is part of the Eastern Cape Province (Rose and Fellingham 1981). And from the early to mid-1900s, maize was widely adopted, replacing sorghum (McCann 2005). From 1996 to 2000, regional EC incidence occurred at an age-standardized rate of 31.2/100,000 for males and 21.8/100,000 females—far higher than next-highest regional cancer incidences at the same time, which were lung cancer in men (6.2/100,000) and cervical cancer in women (19.0/100,000). In South Africa as a whole, the EC rate is 14.3/100,000. Globally, the EC rate is about 4/100,000 for women and 9/100,000 for men. Other hot spots for EC are found in China, Iran, and Zimbabwe (Somdyala et al. 2003).

In response to these developments, the government of South Africa established the first Transkei Oesophagus Cancer Study Group in 1975 to investigate this problem. This Study Group evolved into what is now the Programme on Mycotoxins and Experimental Carcinogenesis (PROMEC) at the Medical Research Council. Through extensive studies, scientists have established a significant correlation between EC and high levels of fumonisin, a mycotoxin produced by a maize fungus (Rheeder et al. 1992).

Evidence suggests that fumonisin exposure causes neural tube defects (NTDs) in human babies by disrupting the uptake of folate in cell lines (Marasas et al. 2004; Missmer et al. 2006). High levels of NTDs were found in a rural district of the Eastern Cape Province (610/100,000) and in rural areas of Limpopo Province (350/100,000), in contrast to about 10/100,000 in urban centers of South Africa (Marasas et al. 2004). High levels of NTDs are also found in those areas of China with high EC and high fumonisin levels. Animal models also link fumonisin with NTDs, but administration of folate partially reduced the mycotoxin's impact (Gelineau–van Waes et al. 2005).

PROMEC has developed and tested a number of strategies to reduce consumers' exposure to mycotoxins. These include food-preparation methods to reduce the mycotoxin (Van der Westhuizen et al. 2010), regulatory standards for maximum allowable exposure (Rheeder et al. 2009), measures for early detection of tumors (Venter 1995), and maize variety selection and breeding programs to develop cultivars with improved host resistance against *Fusarium* ear rot (Rheeder et al. 1990; Rheeder, Marasas, and Van Schalkwyk 1993). This last intervention seemed particularly promising, because such cultivars could be used in rural areas not closely linked to markets or government services.

Transgenic insect-resistant (Bt) maize has emerged as a potential way of reducing fumonisin exposure. Past conventional plant breeding in South Africa has done little to reduce susceptibility of maize to *Fusarium,* the fungus that produces fumonisin. Bt maize contains a gene from the soil bacterium *Bacillus thuringiensis* that encodes for formation of a crystal protein toxic to common lepidopteran maize pests, which includes the maize stalk borer. Insect damage predisposes maize to mycotoxin contamination, because insects create kernel wounds that encourage fungal colonization, and insects themselves may serve as vectors of fungal spores (Sinha 1994; Wicklow 1994; Munkvold, Hellmich, and Rice 1999). Thus, methods that reduce insect damage in maize can also reduce risks of fungal contamination (Wu 2006).

This study examines the extent to which Bt hybrids could reduce mycotoxins at the village level. Although some studies have examined the impact of Bt hybrids on fumonisins on experiment stations both outside and inside South Africa (for example, Munkvold, Hellmich, and Rice 1999; De la Campa et al. 2005; Rheeder et al. 2005), no studies have focused on small farmers' fields. The plan of this study was to sample ears of Bt maize and conventional maize after harvest to measure fungus and mycotoxin levels found on these ears in the regions of the Eastern Cape, where major health problems due to mycotoxins are found, as well as in KwaZulu-Natal, where we were already surveying small farmers about their use of transgenic maize. Unfortunately, in one target region for the study in the Eastern Cape, Bt seed did not arrive in time for planting. In the other Eastern Cape location, the weather was so dry that virtually no maize was planted for several consecutive years. Hence, village-level results reported below are from KwaZulu-Natal alone.

A secondary study objective is to examine whether adoption of Bt maize seed in rural areas could reduce rural consumers' exposure to mycotoxins to levels considered safe by the international health community.

Background

Fumonisins are produced primarily by the fungi *Fusarium verticillioides* and *F. proliferatum*. The former is an almost universal inhabitant of maize (Haliburton and Buck 1986). Fumonisins are a recently discovered class of mycotoxins produced by the fungi *F. verticillioides* (formerly *F. moniliforme*), *F. proliferatum*, and some related species (Rheeder, Marasas, and Vismer 2002). The disease in maize caused by these fungi is called *Fusarium* kernel rot. The first report implicating fumonisins in human disease was connected with high human esophageal cancer rates in Transkei, South Africa, in 1992 (Rheeder et al. 1992). The following year, interest in these mycotoxins increased dramatically after unusually high horse and swine death rates in the United States were linked to contaminated feed (Marasas 1996). Since then, more than 28 fumonisin analogs have been isolated and characterized. Of these, fumonisin B_1 (FB_1) is the most common in maize worldwide (Rheeder, Marasas, and Vismer 2002).

Fumonisin causes toxic effects through its inhibition of ceramide synthase, an enzyme necessary for sphingolipid metabolism (Van der Westhuizen et al. 2008). It is positively associated with EC (Rheeder et al. 1992). Scientists have found similar patterns of high fumonisin concentrations in maize and high-incidence areas of EC in China (Chu and Li 1994). Additional risk factors associated with the development of EC are smoking and alcohol consumption. PROMEC's hypothesis is that in South Africa the increase in EC is due in part to a major shift in cropping patterns from sorghum to maize in these areas and the associated rise in consumer exposure to fumonisins, which are particularly prevalent in maize.

Elevated levels of fumonisin in animal feed cause such diseases as equine leukoencephalomalacia (a disease of the central nervous system) and porcine pulmonary edema, heart failure, and liver damage in swine (Marasas et al. 1988; Kellerman et al. 1990; Ross et al. 1992; Wilson et al. 1992; US Food and Drug Administration 2001). Horses have been shown to exhibit neurological symptoms suggestive of antemortem toxicity after feeding on grain containing 10.56 milligrams/kilogram total fumonisins for 92–122 days, and swine have exhibited liver injury at 23 milligrams/kilogram total fumonisins for 14 days (US Food and Drug Administration 2001). Fumonisin has been shown to cause liver and kidney cancer in rats and liver cancer in mice (Gelderblom et al. 1991; Howard et al. 2001), as well as alterations in kidney function (Bondy et al. 1995; Voss et al. 1995). It is cytotoxic to turkey lymphocytes (Dombrink-Kurtzman et al. 1993).

The Joint United Nations Food and Agriculture Organization/World Health Organization Expert Committee on Food Additives (JECFA) evaluated fumonisins and allocated a group provisional maximum tolerable daily intake of 2 micrograms/kilogram body weight/day for FB_1, FB_2, and FB_3, alone or in combination (Bolger et al. 2001). The International Agency for Research on Cancer has classified FB_1 as a Group 2B carcinogen, that is, possibly carcinogenic to humans (IARC 2002).

Maize flour is the basic starch in South Africa and makes up a large part of all food consumed by people with limited resources. PROMEC's recent survey of consumption patterns in Bizana and Centane in the Eastern Cape Province (Table 2.1) found that adults consumed from 335 to 483 grams of maize/person/day, or between 133 and 176 kilograms per year. In addition, the average consumer drank 32 milliliters/day of beer, and the average self-reported drinker had 1,048 milliliters/day. Beer is primarily made from moldy maize in Bizana and Centane, and hence has high fumonisin levels (Shephard et al. 2005).

The amount of fumonisin in the grain or beer in these regions is very high (Table 2.2). Fumonisin contamination levels are similar for grain

TABLE 2.1   Maize and beer consumption and fumonisin exposure (probable daily intake)

Demographic group	Regions of Transkei (Eastern Cape)	Consumption	Fumonisin exposure (mg/kg body weight/day)	
			Village maize	Commercial maize flour[a]
Maize consumption (g/person/day = ppb)				
Children 1–9 years	Bizana	244	6.60	3.3
	Centane	248	14.14	3.54
Adolescents 10–17 years	Bizana	370	4.05	2.03
	Centane	365	8.33	2.08
Adults 18–65 years	Bizana (women)	335	3.03	1.51
	Bizana (men)	423	3.82	1.91
	Centane (women)	428	8.15	2.04
	Centane (men)	483	9.19	2.23
Beer consumption (mL/person/day)				
Entire population	Centane/Bizana	32	0.2	—
Drinkers only	Centane/Bizana	1,048	6.5	—

Sources: Maize consumption columns 1–4 from Shephard (2006). Beer consumption from Shephard et al. (2005).
Note: — = not applicable.
[a]Calculated from Shephard's data assuming that commercial flour has 200 mg/kg fumonisins.

TABLE 2.2   Fumonisin (FB$_1$) levels in maize and maize products in South Africa

Type of maize sample	Type of producer	Other characteristics	Mean (mg/g)	Range (mg/g)
Maize at harvest	Commercial growers	White grain, 1989	0.57	<0.05–1.12
	Small farmers	Good maize,[a] low prevalence of EC area, 1989	0.67	0.00–3.31
	Small farmers	Moldy maize, low prevalence of EC area, 1989	4.05	0.11–11.34
	Small farmers	Good maize, high prevalence of EC area, 1989	1.84	0.00–5.38
	Small farmers	Moldy maize, high prevalence of EC area, 1989	13.68	3.02–117.50
Maize-based food (village processing)[b]	Small farmers	Good maize in low and high prevalence of EC areas, 1989	0.67–1.84	0.00–5.38
Maize-based food (industrial milling)				
Maize meal	Commercial growers	1991	0.20	0.00–3.90
		1993	0.29	0.00–2.85
Maize grits	Commercial growers	1991	0.13	0.00–0.74
		1993	0.14	0.00–1.38
Animal feed				
Maize bran	Commercial growers	1990–1991	0.90	0.00–4.48
Maize feed	Commercial growers	1993	0.57	0.00–8.55
Maize screenings	Commercial growers	1991–1992	2.10	0.47–4.34

Source: Shephard et al. (1996).
Note: The unit mg/g is equivalent to ppm. EC = esophageal cancer.
[a]Good maize means no obvious mold on maize kernels.
[b]The maize meal and grits produced in the village will have the same levels of fumonisin as the grain.

from large commercial farmers and good-quality grain of smallholder semi-subsistence farmers in low-EC areas (first and second rows in Table 2.2). However, fumonisin is much higher in grain milled in local mills than it is in maize meal or grits milled in commercial mills (rows labeled maize-based food in Table 2.2). Fumonisins are much lower in flour from commercial mills, because the commercial milling process removes the grain's outer shell containing most of the fumonisins.

The last two columns in Table 2.1 show how much fumonisins rural consumers eat. The penultimate column assumes that fumonisin levels are those for home-grown maize in the two regions and that all maize eaten is home grown. In recent years, these assumptions are clearly incorrect, as drought conditions and other factors in some regions have severely limited local maize production in Centane. Whatever consumers are eating is from local stores and is not home grown. If we assume that the fumonisin level in flour is about 200 micrograms/kilogram, which is the level found in commercially milled maize flour, then the amounts consumed would be those shown in the last column of Table 2.1.

Using the levels of fumonisin concentration from maize produced and consumed in villages, the average exposure in all groups is far above the 2 micrograms/kilogram body weight/day that JECFA recommends as the safe upper limit of daily consumption. If all maize flour comes from commercial mills, the mean consumption by children is still 50 percent higher than the JECFA limit. Most other groups (last column in Table 2.1) consume fumonisin at about the JECFA limit, which implies that about half of consumers still take in more than the recommended amount. In South Africa, the Health Department, Medical Research Council, and other experts think that an upper limit of 1 microgram/kilogram body weight/day is a safer level under South African conditions. Most people in these areas consume more than this more restrictive limit.

PROMEC and other organizations have attempted a number of technology and policy interventions over the years. Table 2.3 lists these interventions and indicates whether they were targeted at EC, NTDs, or other animal health problems. Starting in late 1982, PROMEC has also operated a program to identify and encourage commercialization of fumonisin-resistant maize. If the trait could be worked into local farmers' open pollinated varieties through government or public–private partnerships, farmers would not have to buy the seed every year, and expensive government seed supply systems would not be required.

The PROMEC program tested selected commercial hybrids starting in the late 1980s, growing them in Transkei in trials against the landraces (Rheeder 1995). Some showed some resistance, but they were not adapted to conditions in Transkei. As a result, they could not be adopted directly. Some improved lines were identified and turned over to the Agricultural Research Council (ARC). ARC did use these lines in breeding programs for a while but did not produce anything specifically for fumonisin resistance. One problem was that even if you could select cultivars that reduced the visible fungus, you did not necessarily reduce the fumonisin, because sometimes the fungus is not visible on the kernels (known as asymptomatic infection).

When transgenic plants became a possibility, PROMEC contacted private firms such as Pioneer Hi Bred in the United States, which had a research program to introduce genes to disrupt the fumonisin genes in the kernel that were

TABLE 2.3   Possible government interventions and their potential impact

Intervention	Cancer	Neural tube defects	Animal diseases
Maize technology			
Resistance to ear molds	Possible impact	Possible impact	Possible impact
Insect-resistant varieties	Major impact	Major impact	Possible impact
Knockout genes	Potentially high impact	Potentially high impact	Some impact
Farmers or silos grading grain	Grading can eliminate the worst grain, but fumonisin can be prevalent without visible damage	Grading can eliminate the worst grain, but fumonisin can be prevalent without visible damage	Testing for fumonisin levels by feed mills allows them to reject toxic maize
Home processing	Major impact	Major impact	No impact
Supply commercial maize flour	Major impact	Major impact	No impact
Regulations	Limited impact—depends on consumption of commercial maize	No evidence	Voluntary guidelines seem to have worked
Nutrition supplement	No evidence	Folate appears to be effective	No evidence
Nutrition education	Could have impact	Could have impact	No evidence
Surveillance and treatment	Regular screening of rural people for EC with prompt surgery for those with tumors to reduce the associated high mortality rate	No impact	No evidence

Source: Authors.

responsible for producing the fumonisin. PROMEC then saw in the literature that Bt maize, developed to reduce losses from insect pests, also had significantly lower levels of fumonisin (Munkvold, Hellmich, and Rice 1999; Dowd 2001; Bakan et al. 2002; Hammond et al. 2003). When Monsanto introduced the *Bt* gene in South Africa, PROMEC initiated efforts to collaborate with the company to determine whether this technology could reduce mycotoxins in rural villages. The first test compared Bt hybrids with their isolines on ARC experiment stations, where conditions could be controlled. These were first run in 2001/02, with a second round in 2002/03. Results were mixed. In some years in some locations Bt maize had much lower levels of fumonisins than did conventional maize, but in other years or locations little difference was discernable. However, on average Bt maize had 60 percent less fumonisin than conventional hybrids (Rheeder et al. 2005).

The government intervention, which this chapter examines, is the introduction and distribution of Bt maize seed to subsistence farmers. The

TABLE 2.4   Percentage of smallholder farmers using purchased seed, by region, 2001

| | Mpumalanga | | KwaZulu-Natal Hlabisa | Limpopo Venda | Eastern Cape | |
	Northern Highveld	Southern Highveld			Mqanduli	Flagstaff
Percentage of farmers using purchased seed	78	76	98	81	13	20

Source: Marnus Gouse, unpublished data.

hypothesis is that Bt maize would control stalk borers, thus reducing fungus levels and concomitantly reducing levels of fumonisin in home-grown maize. Hybrid Bt maize seeds have spread widely in South Africa in recent years but are mainly used by commercial farmers rather than subsistence farmers in the Eastern Cape (see Chapter 1 for a discussion of the spread of GM maize in South Africa).

The introduction of Bt maize to small farms in the Eastern Cape is unlikely to take place without some government assistance, because seed companies would not find it profitable to serve smallholder farmers in these areas. The farmers are poor, do not buy seeds annually (see Table 2.4), and rarely use complementary inputs like fertilizer on their soil. Thus, either government extension services must develop Bt hybrids for the region and provide these seeds to farmers, or the government should subsidize private companies to provide these seeds.

Methods and Results of the Rockefeller Study

Maize samples were collected from farmers in three locations in the province of KwaZulu-Natal after harvest. Samples were taken in areas where we were also surveying the economic impact of Bt maize on farmers' incomes. We contacted farmers in our sample who we knew were using Bt maize. In some cases when our contact farmers had already consumed or sold their Bt maize, we contacted their neighbors. Farmers identified the grain as Bt hybrids, non-Bt hybrids, or local varieties. The number of samples varied between 50 and 80 maize samples each year.

The mycological analysis was conducted as follows. Briefly, subsamples of kernels (80–100 grams) were surface sterilized for 1 minute in 3.5 percent commercial sodium hypochlorite solution and rinsed twice in sterile distilled water. One hundred kernels (5 kernels/90 millimeter petri dish) were then transferred to malt extract agar (1.5 percent) containing novobiocin (150 milligrams/liter), and the agar plates were incubated at 25°C in the dark for 5–7 days. All isolated fungi were recorded and identified according to their morphological characteristics on the agar plates.

Fumonisin level was determined as follows. Each sample was ground in a laboratory mill to a fine meal and extracted with methanol:water (3:1) by homogenization. An aliquot was applied to a strong anion exchange solid-phase extraction cartridge, and the fumonisins were eluted with 1 percent acetic acid in methanol. The purified extracts were evaporated to dryness, redissolved in methanol, and derivatized with *o*-phthaldialdehyde. The derivatized extracts were analyzed by reversed-phase high-performance liquid chromatography using a Phenomenex Luna 5: C18(2) column and fluorescence detection.

Statistical analysis was done on the mycology and fumonisin data using SPSS 13 software (Chicago, United States) to determine any significant differences between the groups (that is, traditional, commercial, and Bt maize) of samples at the three locations. Analyses were conducted on the *F. verticillioides* + *F. proliferatum* (FvFp) and total fumonisin variables only, with natural log transformation of the latter variable.

The results of the village surveys (Table 2.5) show a clear advantage of Bt maize over conventional hybrids and traditional maize seed. Bt maize had 40 percent less fumonisin than did traditional cultivars. Relative to the non-Bt commercial maize hybrid, Bt maize had on average 16 percent less fumonisin.

The table also shows variability of fumonisin levels in different years and different locations. These differences may result from weather, other environmental factors, or hybrid characteristics. A recent study (De la Campa et al. 2005) of fumonisin in maize in fields managed by scientists rather than farmers in Argentina and the Philippines found that although the use of Bt maize explains much of the variation in fumonisin levels, location or weather explain more of the variation. In addition, levels of insect damage and use of different hybrids also explain the variation. Our data show similar results: on average Bt maize has less fumonisin, but levels vary because of weather and other factors. As noted in Table 2.5, in the first year in Hlabisa Bt hybrids had traits that attracted birds to eat kernels, which allowed the fungus that causes fumonisin to multiply rapidly and led to high levels of fumonisin.

These results can also be compared to those from South African experiment station studies conducted by PROMEC on ARC research stations Potchefstroom and Vaalharts in 2002 and 2003. These experiment stations are in the heart of the commercial maize-growing area of South Africa. These experiments compared commercial Bt hybrids with their isolines (identical

TABLE 2.5 Comparison of total fumonisin levels in maize in rural KwaZulu-Natal, 2004–2007 (mg/kg = ppb)

Year/location	N^a	Traditional maize	N	Commercial maize[b]	N	Bt maize
2004						
Simdlangentsha	5	753 ± 814	8	623 ± 917	7	239 ± 411
Hlabisa	4	159 ± 91	11	450 ± 627	8	1,147 ± 1,432[c]
2005						
Simdlangentsha	6	271 ± 352	12	815 ± 1147	7	396 ± 395
Hlabisa	2	250 ± 353	15	472 ± 505*[d]	11	22 ± 25*[d]
2006						
Simdlangentsha	5	996 ± 1,290	6	1,200 ± 949	2	152 ± 64
Hlabisa	9	2,065 ± 3,232	20	1,074 ± 2,117	13	804 ± 989
Dumbe	6	426 ± 606	3	2,595 ± 4,140	5	1,280 ± 2,086
2007						
Simdlangentsha		n.a.	14	1,812 ± 3,230	6	51 ± 70
Hlabisa		n.a.	17	348 ± 1,154	10	129 ± 138
Dumbe	6	3,391 ± 5,239	18	848 ± 1,443	4	500 ± 927
Average	43	1,233 ± 2602	124	886 ± 1749	73	748 ± 1518

Source: Surveys by authors.

Note: n.a. = not available.

[a]N is the number of observations in this category.

[b]Commercial maize includes nontransgenic and transgenic hybrids but without Bt genes.

[c]High levels of fumonisin are a result of untimely late rains just before harvest linked with maize ear morphology of the Bt hybrid (which has since been removed from commercial production), which resulted in attacks by birds and fungal damage to the ears.

[d]The means of these variables are significantly different from the other starred mean in the row at the 5 percent level.

hybrids except for the addition of Bt). As in the village studies, there was considerable diversity in results among different hybrids, different locations, and different years. The authors of this study (Rheeder et al. 2005, S88) concluded that "the incidence of fumonisin-producing *Fusarium* species, stalk borers (*B. fusca*), and levels of fumonisin, were generally lower in the Bt hybrids." The fumonisin levels in the Bt hybrids were on average 51 percent lower than the non-Bt isoline at Potchefstroom in 2002, 83 percent lower at Potchefstroom in 2003, and 39 percent lower in Vaalharts in 2003.

Implications—How Much Could This Technology Reduce Mycotoxin Exposure?

If all farmers in the Eastern Cape shifted to Bt maize and obtained the same results as farmers in our sample in KwaZulu-Natal, would it reduce their intake of fumonisin to safe levels? To test this possibility, we simulated the impact of reducing fumonisin levels by 62 percent using the data introduced in Table 2.1.

If Bt maize could be introduced into this area and adopted by 100 percent of farmers and if the reduction in fumonisin exposure due to Bt were the same as that found in the maize from KwaZulu-Natal villages (column 4 of Table 2.5), then there would be a dramatic reduction in fumonisin exposure (see Table 2.6). In Bizana, most groups of people except children would have lower exposure than the suggested maximum level of 2 micrograms/kilogram body weight/day. In Centane, however, all groups would still remain above the suggested level.

In fact, the combination of the declining production of maize in the area (which has forced families to eat more commercially milled maize) and a switch to Bt maize by those who continue to grow maize could reduce exposure dramatically. The only obvious way to reduce consumption to the level of 2 micrograms/kilogram body weight/day is by stopping consumption of home-grown maize (see last column of Table 2.1) or through farmers' adoption of a cultivar that completely disrupts the production of fumonisin in maize. Cultivars of this type are being developed in research programs at the

TABLE 2.6   Fumonisin exposure in the Eastern Cape with Bt maize adoption

| Demographic group | Region | Fumonisin exposure (mg/kg = ppb body weight/day) | |
		Village maize (baseline)	Simulation 1 (Bt maize adoption)[a]
Maize consumption			
Children 1–9 years	Bizana	6.60	2.51
	Centane	14.14	5.37
Adolescents 10–17 years	Bizana	4.05	1.54
	Centane	8.33	3.17
Adults 18–65 years	Bizana (women)	3.03	1.15
	Bizana (men)	3.82	1.45
	Centane (women)	8.15	3.10
	Centane (men)	9.19	3.49
Beer consumption			
Entire population	Centane/Bizana	0.2	0.2[a]
Drinkers only	Centane/Bizana	6.5	6.5

Sources: First three columns are from Shephard (2006). Fourth column is calculated by the authors.

Note: Simulation 1 assumes that all village maize is Bt and has 62 percent less fumonisin that for conventional maize.

[a] Exposure by consumption of beer stays the same, because people choose fungus-infected (moldy) grain to make beer.

Danforth Center in St. Louis, MO, United States, but they are not in field trials in the United States or South Africa.

Other village-level options that PROMEC has explored include improved sorting and washing of maize kernels to remove fumonisin (Van der Westhuizen et al. 2010), or new cooking techniques to reduce the consumption of fumonisin. Washing grain for 10 minutes and grading it appear to be particularly promising, reducing fumonisin in cooked food by 65 percent (Van der Westhuizen et al. 2010). The challenge with this option is getting the word out to resource-poor households and encouraging them to use this simple food-preparation procedure.

Another possible intervention would be to supply more commercial grain, with lower fumonisin levels, to farmers through health and welfare programs. As Table 2.2 shows, shifting from local grain to commercially milled grain would substantially reduce exposure, but there could be a loss of consumer benefits if commercially milled grain is more expensive and consumers prefer the taste of their own grain.

Stricter regulations to ban the sale of grain with more than 1 ppm fumonisin might have an impact on urban consumers who buy maize in markets but would not affect rural consumers who do not buy maize meal. To prevent NTDs, the government is requiring that all grain be fortified with folate. However, subsistence farmers cannot easily access folate supplements, and their diets do not naturally contain high folate levels. Finally, brush biopsies can be used for early detection of EC (Venter 1995). If detected early enough, tumor removal is usually a successful operation, but this is clearly the intervention of last resort.

Conclusions and Future Research

This study provides evidence that the adoption of Bt maize can reduce exposure of subsistence farmers in South Africa to the mycotoxin fumonisin. The spread of Bt maize could ameliorate but not solve the problem of fumonisin in human and animal diets. However, getting Bt maize seed to small farmers in the Eastern Cape would be a major undertaking. Farmers are not using hybrids in these areas. They are instead using relatively well-adapted open pollinated varieties, and they save their own seed. Even with the good will of a major seed company, we were not able to get seed of existing Bt hybrids out to farmers so that they could test it. To establish widespread acceptance of Bt cultivars, some organization would have to either breed open pollinated varieties that are well adapted to this area or provide well-adapted hybrids every year or two. ARC has participatory breeding programs in some areas of South Africa that could be used for this purpose, but so far ARC is not doing participatory breeding for reduced fumonisin or using Bt lines. The extension system is very limited in what it can do. Farmers in some of these areas have almost given up planting maize.

Changes in eating habits to use more commercial maize meal and the increasing use of rice and vegetables have already helped reduce fumonisin exposure. These changes seem to be due to a combination of higher welfare payments (which may have allowed some families to buy more grain) and also to poor rainfall (which has led to lower production of maize). None of these were policies to improve diets and when taken in combination with urbanization (which has increased the availability and status of "fast foods"), the diets of the poor may well have reduced fumonisin levels but at the same time become considerably less healthy.

Currently, the government of South Africa and the provincial governments clearly have their hands full with HIV/AIDS. However, if the government does decide to attack EC, a thorough evaluation of the efficacy of the various options available to reduce exposure to fumonisin is needed, and then a serious study of the costs of the different types of interventions must be conducted.

References

Bakan, B., D. Melcion, D. Richard-Molard, and B. Cahagnier. 2002. "Fungal Growth and *Fusarium* Mycotoxin Content in Isogenic Traditional Maize and Genetically Modified Maize Grown in France and Spain." *Journal of Agricultural and Food Chemistry* 50 (4): 728–731.

Bolger, M., R. D. Coker, M. DiNovi, D. Gaylor, W. Gelderblom, M. Olsen, N. Paster, et al. 2001. "Fumonisins." In *Safety Evaluation of Certain Mycotoxins in Food.* WHO Food Additives Series 47, FAO Food and Nutrition Paper 74. Prepared by the 56th Meeting of the Joint FAO/WHO Expert Committee on Food Additives (JECFA), Geneva.

Bondy, G., C. Suzuki, M. Barker, C. Armstrong, S. Fernie, L. Hierlihy, P. Rowsell, and R. Mueller. 1995. "Toxicity of Fumonisin B_1 Administered Intraperitoneally to Male Sprague-Dawley Rats." *Food and Chemical Toxicology* 33: 653–665.

Chu, F. S., and G. Y. Li. 1994. "Simultaneous Occurrence of Fumonisin B_1 and Other Mycotoxins in Moldy Corn Collected in the People's Republic of China in Regions with High Incidences of Esophageal Cancer." *Applied and Environmental Microbiology* 60: 847–852.

De la Campa, R., D. C. Hooker, J. D. Miller, A. W. Schaafsma, and B. G. Hammond. 2005. "Modeling Effects of Environment, Insect Damage, and Bt Genotypes on Fumonisin Accumulation in Maize in Argentina and the Philippines." *Mycopathologia* 159: 539–552.

Dombrink-Kurtzman M. A., T. Javed, G. A. Bennett, J. L. Richard, L. M. Cote, and W. B. Buck. 1993. "Lymphocyte Cytotoxicity and Erythrocyte Abnormalities Induced in Broiler Chicks by Fumonisins B_1 and B_2 and Moniliformin from *Fusarium proliferatum*." *Mycopathologia* 124: 47–54.

Dowd, P. F. 2001. "Biotic and Abiotic Factors Limiting Efficacy of Bt Corn in Indirectly Reducing Mycotoxin Levels in Commercial Fields." *Journal of Economic Entomology* 94 (5): 1067–1074.

Gelderblom, W.C.A., N.P.J. Kriek, W.F.O. Marasas, and P. G. Thiel. 1991. "Toxicity and Carcino-genicity of the *Fusarium moniliforme* Metabolite Fumonisin B_1, in Rats." *Carcinogenesis* 12: 1247–1251.

Gelineau–van Waes, J., L. Starr, J. Maddox, F. Alleman, K. A. Voss, J. Wilberding, and R. T. Riley. 2005. "Maternal Fumonisin Exposure and Risk for Neural Tube Defects: Mechanisms in an in vivo Mouse Model." *Birth Defects Research (Part A): Clinical and Molecular Teratology* 73: 487–497.

Haliburton, J. C., and W. B. Buck. 1986. "Equine Leukoencephalomalacia: An Historical Review." *Current Topics in Veterinary Medicine and Animal Science* 33: 75–79.

Hammond, B., K. Campbell, C. Pilcher, A. Robinson, D. Melcion, B. Cahagnier, J. Richard, et al. 2003. "Reduction of Fumonisin Mycotoxins in Bt Corn." *Toxicologist* 72 (S-1): abstract 1217.

Howard, P. C., R. M. Eppley, M. E. Stack, A. Warbritton, K. A. Voss, R. J. Lorentzen, R. M. Kovach, et al. 2001. "Fumonisin B_1 Carcinogenicity in a Two-Year Feeding Study Using F344 Rats and $B6C3F_1$ Mice." *Environmental Health Perspectives* 109 (Suppl 2): 277–282.

IARC (International Agency for Research on Cancer). 2002. "Some Traditional Herbal Medicines, Some Mycotoxins, Naphthalene and Styrene." In *Monographs on the Evaluation of Carcinogenic Risks to Humans,* 301–366. Lyon, France: International Agency for Research on Cancer Press.

Kellerman, T. S., W.F.O. Marasas, P. G. Thiel, W.C.A. Gelderblom, M. Cawood, and J.A.W. Coetzer. 1990. "Leukoencephalomalacia in Two Horses Induced by Oral Dosing of Fumonisin B1." *Onderstepoort Journal of Veterinary Research* 57: 269–275.

Marasas, W.F.O. 1996. "Fumonisins: History, World-Wide Occurrence and Impact." In *Fumonisins in Food,* edited by L. S. Jackson, J. W. De Vries, and L. B. Bullerman, 1–17. New York: Plenum.

Marasas, W.F.O., T. S. Kellerman, W.C.A. Gelderblom, J.A.W. Coetzer, P. G. Thiel, and J. J. van der Lugt. 1988. "Leukoencephalomalacia in a Horse Induced by Fumonisin B_1, Isolated from *Fusarium moniliforme*." *Onderstepoort Journal of Veterinary Research* 55: 197–203.

Marasas, W.F.O., R. L. Riley, K. A. Hendricks, V. L. Stevens, T. W. Sadler, J. Gelineau–van Waes, S. A. Missmer, et al. 2004. "Fumonisins Disrupt Sphingolipid Metabolism, Folate Transport, and Neural Tube Development in Embryo Culture and *in vivo*: A Potential Risk Factor for Human Neural Tube Defects among Populations Consuming Fumonisin-Contaminated Maize." *Journal of Nutrition* 134: 711–716.

McCann, J. C. 2005. *Maize and Grace: Africa's Encounter with a New World Crop, 1500–2000.* Cambridge, MA, US: Harvard University Press.

Missmer, S. A., L. Suarez, M. Felkner, E. Wang, A. H. Merrill Jr., K. J. Rothman, and K. A. Hendricks. 2006. "Exposure to Fumonisins and the Occurrence of Neural Tube Defects along the Texas-Mexico Border." *Environmental Health Perspectives* 114: 237–241.

Munkvold, G. P., R. L. Hellmich, and L. G. Rice. 1999. "Comparison of Fumonisin Concentrations in Kernels of Transgenic Bt Maize Hybrids and Nontransgenic Hybrids." *Plant Disease* 83 (2): 130–138.

Rheeder, J. P. 1995. "Study of the Factors Associated with the Incidence of *Fusarium moniliforme* Ear Rot of Maize in South Africa and Transkei." PhD Thesis, University of the Orange Free State, Bloemfontein, South Africa.

Rheeder, J. P., W.F.O. Marasas, and D. L. Van Schalkwyk. 1993. "Incidence of *Fusarium* and *Diplodia* Species in Naturally Infected Grain of South African Maize Cultivars: A Follow-up Study." *Phytophylactica* 25: 43–48.

Rheeder, J. P., W.F.O. Marasas, and H. F. Vismer. 2002. "Production of Fumonisin Analogs by *Fusarium* Species." *Applied Environmental Microbiology* 68 (5): 2101–2105.

Rheeder, J. P., W.F.O. Marasas, P. S. Van Wyk, and D. J. Van Schalkwyk. 1990. "Reaction of South African Maize Cultivars to Ear Inoculation with *Fusarium moniliforme, F. graminearum* and *Diplodia maydis.*" *Phytophylactica* 22: 213–218.

Rheeder, J. P., W.F.O Marasas, P. G. Thiel, E. W. Sydenham, G. S. Shephard, and D. L. Van Schalkwyk. 1992. "*Fusarium moniliforme* and Fumonisins in Corn in Relation to Human Esophageal Cancer in Transkei." *Phytopathology* 82: 353–357.

Rheeder, J. P., H. F. Vismer, L. van der Westhuizen, G. Imrie, P. Gatyeni, D. Thomas, G. Shephard, et al. 2005. "Effect of Bt Corn Hybrids on Insect Damage, Incidence of Fumonisin-Producing *Fusarium* Species and Fumonisin Levels in South Africa." *Phytopathology* 95 (supplement): S88.

Rheeder, J. P., G. S. Shephard, H. F. Vismer, and W.C.A. Gelderblom. 2009. "Guidelines on Mycotoxin Control in South African Foodstuffs: From the Application of the Hazard Analysis and Critical Control Point (HACCP) System to New National Mycotoxin Regulations." Medical Research Council Policy Brief. Accessed October 2009. www.mrc.ac.za/policybriefs/policybriefs.htm.

Rose, E. F., and S. A. Fellingham. 1981. "Cancer Patterns in Transkei." *South African Journal of Science* 77: 555–561.

Ross, P. F., L. G. Rice, G. D. Osweiler, P. E. Nelson, J. L. Richard, and T. M. Wilson. 1992. "A Review and Update of Animal Toxicoses Associated with Fumonisin-Contaminated Feeds and Production of Fumonisins by *Fusarium* Isolates." *Mycopathologia* 117: 109–114.

Shephard, G. S. 2006. "Mycotoxins in the Context of Food Risks and Nutrition Issues." In *The Mycotoxin Factbook: Food and Feed Topics,* edited by D. Barug, D. Bhatnagar, H. P. van Egmond, J. W. van der Kamp, W. A. van Osenbruggen, and A. Visconti, 21–36. Wageningen, the Netherlands: Wageningen Academic Press.

Shephard, G. S., P. G. Thiel, S. Stockenstrom, and E. W. Sydenham. 1996. "Worldwide Survey of Fumonisin Contamination of Corn and Corn-Based Products." *Journal of AOAC International* 79 (3): 671–687.

Shephard G. S., L. van der Westhuizen, P. M. Gatyeni, N.I.M. Somdyala, H.-M. Burger, and W.F.O. Marasas. 2005. "Fumonisin Mycotoxins in Traditional Xhosa Maize Beer in South Africa." *Journal of Agricultural and Food Chemistry* 53: 9634–9637.

Sinha, A. K. 1994. "The Impact of Insect Pests on Aflatoxin Contamination of Stored Wheat and Maize." In *Stored Product Protection: Proceedings of the 6th International Working Conference on Stored-Product Protection,* edited by E. Highley, E. J. Wright, H. J. Banks, and B. R. Champ, 1059–1063. Wallingford, UK: CAB International.

Somdyala, N.I.M., W.F.O Marasas, F. S. Venter, H. F. Vismer, W.C.A. Gelderblom, and S. A. Swanevelder. 2003. "Cancer Patterns in Four Districts of the Transkei Region—1991–1995." *South African Medical Journal* 93: 144–148.

US Food and Drug Administration. 2001. *Guidance for Industry. Fumonisin Levels in Human Foods and Animal Feeds.* Accessed November 9, 2001. www.fda.gov/Food/GuidanceCompliance RegulatoryInformation/GuidanceDocuments/ChemicalContaminantsandPesticides/ ucm109231.htm.

Van der Westhuizen, L., G. S. Shephard, J. P. Rheeder, N.I.M. Somdyala, and W.F.O Marasas. 2008. "Sphingoid Base Levels in Humans Consuming Fumonisin-Contaminated Maize from Low and High Oesophageal Cancer Incidence Areas: A Cross-Sectional Study." *Food Additives and Contaminants* 25 (11): 1385–1391.

Van der Westhuizen, L., G. S. Shephard, J. P. Rheeder, H. M. Burger, W.C.A. Gelderblom, C. P. Wild, and Y. Y. Gong. 2010. "Simple Intervention Method to Reduce Fumonisin Exposure in a Subsistence Maize-Farming Community in South Africa." *Food Additives and Contaminants* 27: 1582–1588.

Venter, F. S. 1995. "Standardized Methodology for Cytotechnology with the Nabeya Oesophageal Brush Biopsy Capsule." *Medical Technology SA* 9: 153–154.

Voss, K. A., W. J. Chamberlain, C. W. Bacon, R. A. Herbert, D. B. Walters, and W. P. Norred. 1995. "Subchronic Feeding Study of the Mycotoxin Fumonisin B_1 in B6C3F1 Mice and Fischer 344 Rats." *Fundamental and Applied Toxicology* 24: 102–110.

Wicklow, D. T. 1994. "Preharvest Origins of Toxigenic Fungi in Stored Grain." In *Stored Product Protection: Proceedings of the 6th International Working Conference on Stored-Product Protection,* edited by E. Highley, E. J. Wright, H. J. Banks, and B. R. Champ, 1075–1081. Wallingford, UK: CAB International.

Wilson, T. M., P. F. Ross, D. L. Owens, L. G. Rice, S. A. Green, S. J. Jenkins, and H. A. Nelson. 1992. "Experimental Reproduction of ELEM. A Study to Determine the Minimum Toxic Dose in Ponies." *Mycopathologia* 117: 115–120.

Wu, F. 2006. "Mycotoxin Reduction in Bt Corn: Potential Economic, Health and Regulatory Impacts." *Transgenic Research* 15: 277–289.

Chapter 3

Genetically Modified Cotton in Uganda: An Ex Ante Evaluation

Daniela Horna, Patricia Zambrano, José Falck-Zepeda,
Theresa Sengooba, and Miriam Kyotalimye

The Ugandan government has recognized the need to increase the performance of cotton and the potential of crop biotechnologies, particularly the role of genetically modified (GM) varieties to improve cotton production and thus the economy in general. In 2008, the National Biosafety Committee of Uganda approved the guidelines for implementing confined trials, which enabled testing the environmental safety and performance of insect-resistant (Bt) and herbicide-tolerant (HT) cotton varieties. The implementation of the confined trials started in May 2009.

The release into the environment and eventual commercialization of GM cotton seed may raise questions about the socioeconomic impact of this technology for Ugandan farmers. The economic impact of GM technologies can be assessed at different levels. The objective of this study is to provide an ex ante evaluation of the potential impact of GM cotton adoption in Uganda at the farm level.[1] To understand the context of this evaluation, we briefly review the literature available on ex ante evaluations of GM crops in Africa south of the Sahara (SSA). Next we describe the cotton value chain in Uganda and explain the main production systems. In the third section, we present the methodological approach used in the study to compare the profitability of the different cotton production alternatives. Finally, we present our main findings, conclusions, and policy recommendations and examine further considerations for evaluating the potential impact of GM cotton in Uganda.

---

1 This study was part of a multilevel evaluation carried out in 2007 with the support of the Program for Biosafety Systems. A full report of the project findings can be found in the forthcoming IFPRI book *Socioeconomic Considerations in Biosafety Decisionmaking: Methods and Implementation.*

Ex Ante Impact of GM Crops in SSA

The adoption of GM varieties worldwide has expanded considerably since they were first commercialized in 1996. Originally only four countries approved the adoption and use of transgenic crops covering an area of just above 11 million hectares. In the case of cotton, GM varieties have been commercialized in 10 countries: Argentina, Australia, Brazil, China, Colombia, India, Mexico, South Africa, the United States, and most recently, Burkina Faso. In parallel, the areas of non-GM conventional cotton have been decreasing in favor of Bt and HT cotton. James (2011) estimates that in 2011, more than 68 percent of world cotton was planted to GM varieties, including both open-pollinated varieties and hybrids.

In 2009 the International Food Policy Research Institute published a review of the economic impact of these crops in developing economies and the methods used to evaluate this impact (Smale et al. 2009). The review revealed that very few studies have addressed the potential or actual impact of GM crops on smallholders in SSA. Aside from numerous publications of the actual impact of Bt cotton and Bt maize in South Africa, the number of ex ante evaluations is still limited in Africa. In East Africa De Groote et al. (2003) evaluated the potential for Bt maize; Edmeades and Smale (2006) discussed the potential impact of a GM banana on smallholder farmers in the Uganda highlands. In West Africa, Horna et al. (2008) assessed ex ante GM vegetables in Ghana. More recently Vitale at al. (2008, 2010) have documented the impact of Bt cotton in Burkina Faso.

Given the importance of cotton in SSA and the availability of the technology, most of the ex ante evaluations have concentrated on GM cotton, especially Bt cotton. In West Africa for instance, Cabanilla, Abdoulaye, and Sanders (2005) developed a linear-programming model to assess the potential cost to West Africa, Mali in particular, of not adopting Bt cotton. Elbehri and MacDonald (2004) and Langyintuo and Lowenberg-DeBoer (2006) have carried out evaluations mainly based on trade models. Most recently, a study by Falck-Zepeda, Horna, and Smale (2008) examined the potential of Bt cotton for the West Africa region with emphasis on Benin, Burkina Faso, Mali, Senegal, and Togo using an economic surplus approach.

These studies conclude that there would be significant losses in economic benefits if Bt cotton is not adopted. Several of these studies also point to important institutional issues as significant determinants of the level and social distribution of benefits. A key institutional issue seems to be the decision about the technology development, whether it will be imported, adapted, or generally adopted (Cabanilla, Abdoulaye, and Sanders 2005; Falck-Zepeda, Horna, and Smale 2008). The technology fee (price charged to farmers) is also a cross-cutting issue that will determine the potential benefits of GM technologies in developing economies (Smale et al. 2009). Another critical issue is the ability of the innovator or the technology-transfer agent to transmit to farmers the necessary knowledge to manage the technology under field conditions. As concluded by Falck-Zepeda et al. (2008), GM technologies need to be part of a broad, integrated pest-management—or better yet, an integrated crop-management—strategy for implementation in SSA.

The Ugandan Cotton Sector

The history of cotton in Uganda is well documented in the literature (You and Chamberlin 2004; Cotton Development Organisation 2006; Baffes 2009; Tschirley, Poulton, and Labaste 2009). The long tradition of cotton cultivation gives the crop a historic significance. Cotton was introduced to the central region of Uganda in about 1903, and over the years it has spread to the rest of the country. In the 1950s cotton and coffee became— and up to 1970 they remained—the most important sources of revenue for the government.

Several studies have noted that cotton has the potential to improve the welfare of about 250,000 low-income farming households (Gordon and Goodland 2000; Baffes 2009). At the aggregate national level, however, cotton constitutes just 2.2 percent of all crop production and thus plays a relatively small part in rural livelihoods (You and Chamberlin 2004). From a trade point of view, cotton ranks third among agricultural commodities exported, although it only accounts for 2–5 percent of Ugandan total exports (Serunjogi et al. 2001). In 2007 cotton exports were valued at 36 million US dollars with Switzerland, United Arab Emirates, Singapore, and the United Kingdom generating more than 80 percent of these revenues (ICAC 2008; FAO 2010). All organic cotton is exported to the EU, mainly Switzerland. The local industry consumes approximately 7–10 percent of the lint produced (ACE 2006; Uganda 2007).

Cotton is widely cultivated in more than 30 districts across the country (see Figure 3.1) because of the favorable agroclimatic conditions (You and Chamberlin 2004). The most important producing areas are located in the northern and western regions of Uganda.

FIGURE 3.1   Cotton-growing regions in Uganda

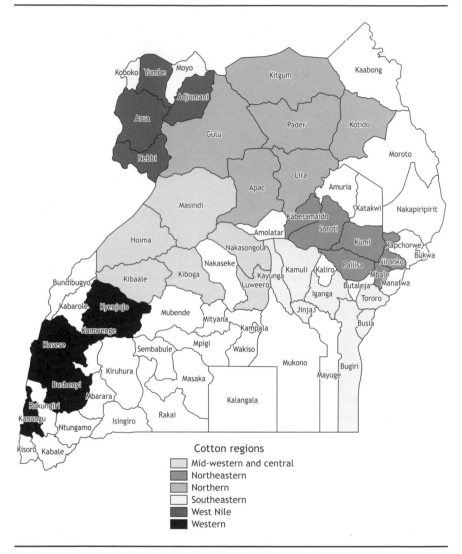

Source: Map by Z. Guo (International Food Policy Research Institute)

Cotton Value Chain

The cotton sector in Uganda is characterized by vertical integration (Figure 3.2). Understanding the production systems and the value chain can help identify potential agronomic and institutional constraints on the adoption and introduction of GM technology in Uganda.

FIGURE 3.2   Cotton value chain in Uganda

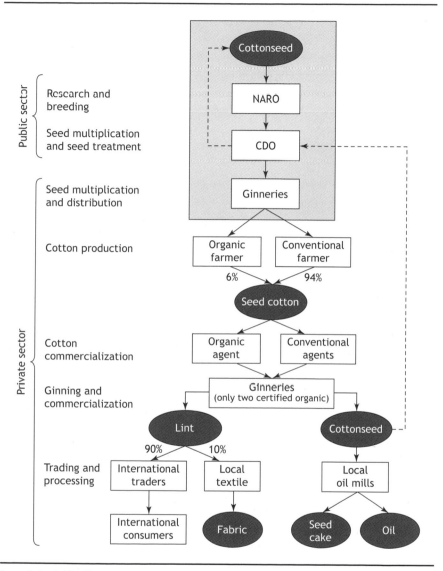

Source: Authors' elaboration.

Note: CDO = Cotton Development Organisation; NARO = National Agricultural Research Organisation.

Seed Value Chain

The cotton value chain depends on the availability and the quality of seeds. Ugandan production is characterized by the use of only one seed variety available for both conventional and organic planting, the Bukalasa Pedigree Albar. The one-seed policy has been promoted as a way to guarantee quality homogenization. From the postharvest and marketing perspective, this is an advantage, because a single variety ensures uniformity in the production of lint and yarn. The dependence on a single variety, however, increases the vulnerability to pests and diseases and represents a potential risk.

The main actors in the cotton seed chain in Uganda are the cotton producers, the National Agricultural Research Organisation (NARO) of Uganda, the Cotton Development Organisation (CDO), and some private ginneries. NARO[2] coordinates and oversees all aspects of agricultural research in Uganda; it is in charge of cotton research breeding and technology development. According to ACE (2006), its breeding work had led to increased yields (from 740–1,230 kilograms/hectare to 3,700–5,000 kilograms/hectare currently) and enhanced ginning turnout (from 33 percent up to 37 percent in areas of high fertility).

However, the multiplication and seed-distribution process needs more attention. The basic cotton seed is developed by NARO but multiplied by CDO and mainly distributed to the ginneries for commercial multiplication and distribution to farmers. The seed that farmers use is entirely channeled through the ginneries, NARO, and CDO. The CDO[3] regulates, coordinates, and promotes all aspects of the cotton subsector in Uganda. CDO also monitors cotton production and marketing and provides policy advice regarding the crop (Cotton Development Organisation 2006). CDO is therefore responsible for delinting, grading, and dressing all seed that will be given to ginneries for distribution. The seed is distributed free of charge to farmers.[4]

Availability of cotton seed is a limiting constraint for improving cotton productivity in Uganda. The need for improved varieties and certified seed is probably the most important constraint encountered in cotton production (Serunjogi et al. 2001). According to the Uganda Export Promotion Board (UEPB 2007), cotton exports in 2006 fell 29 percent compared to the

---

2   NARO was established by an act of Parliament on November 21, 2005 (www.naro.go.ug/About%20NARO/aboutnaro.htm).

3   An act of Parliament established CDO in 1994 (http://cdouga.org/).

4   In our estimations we assumed a zero value for seed, as it is distributed free of charge to the farmers. At harvest, farmers do pay for ginning and transportation services.

previous year. Among other factors, the low performance in 2006 was related to late planting, but mainly to the use of ungraded fuzzy (not delineated) seed, leading to high seed wastage and increased cost of provision of planting seed.

Given the existing limitations in the seed value chain, the introduction of GM seed in this system needs to be carefully thought out. First, the one-variety policy would probably need to be changed. The possibility exists that the developers may not introduce the GM trait into the local variety, as it may not be part of their marketing and product diffusion strategy. Alternatively, developers may have improved varieties suited to the area. The participation of the public and private sectors in seed multiplication and distribution has to be discussed up-front. Would the technology developer be in charge of seed distribution? Is the government interested in negotiating a good price for cotton producers?

The technology fee is another critical aspect that needs to be carefully discussed. The development of GM seed for Uganda represents a signifi-cant investment that must be paid for. It is quite likely that GM seed will not be delivered free of charge, as has been the case in most adopting countries. It is possible that the government of Uganda would assume part of the cost, but it can also be expected that farmers will need to pay for the technology. For resource-poor farmers the big incentive to adopt a technology is often a guaranteed profit, especially for a cash crop like cotton.

Product Value Chain

Farmers, intermediate agents, and ginners/exporters are the main actors in the product value chain. As mentioned above, farmers obtain the seed from the ginners. Ginners traditionally provide farmers with fertilizers and pesticides. Farmers pay back at harvest with either cotton or cash. The level of inputs used is still limited, however, and often farmers decide to plant cotton only because of the secure market and fixed price. In the northern part of the country there are few alternatives to cotton, either as a single crop or as part of a rotation.

At harvest, farmers can take their production to the ginneries, but often the volume produced is so small it does not justify paying for transport. Most com-monly, intermediary agents gather the production of several farmers and take it to the ginneries. Intermediary agents can either work for the ginnery or be independent. There are more than 50 ginneries distributed all over the coun-try. Thirty-one different companies privately own the ginneries. Given the irreg-ular cotton production, ginneries compete for access to cotton areas, as most ginneries operate below their potential capacity, although the operating gin-ning machines are of poor quality. In most cases, ginneries are active only during the harvest period. Although cotton production in Uganda does not cover the ginning potential and most companies work with excess capacity, most machines are rather old, and the quality of the turnout is low. An increase in overall cotton production can favor individual actors (producers, ginners, and government) and thus the economy in general. If the introduction of GM cotton boosted cotton yields, as expected, ginnery operations would therefore increase.

The challenging part from the standpoint of marketing channels will be devising ways to keep GM cotton from mixing with conventional varieties, or even worse, with organic cotton. However, in Uganda organic lint is already separated from conventional lint by using different ginneries for the two types. After separation of cotton seed from cotton lint, organic and conventional seeds are pooled together. The recycled seeds go back to ginneries, which deliver them to the farmers (see Figure 3.2). Under the current organization of the seed value chain, this might not be a problem, as both the organic producers and the biotech companies have clear incentives to separate GM seed from other seeds. For producers and ginneries, it would be important to keep seeds used for organic production from mixing with GM seeds, and biotech companies would avoid free recycling of the seed.

Cotton Production Systems

TRADITIONAL SYSTEM

Cotton in Uganda is grown under rainfed conditions largely by smallholders (Gordon and Goodland 2000). As a rainfed system, cotton production is dependent on rainfall patterns. Climatic events, especially variability in precipitation, can severely affect cotton yields. The timeliness of the precipitation can determine a good or bad year for the crop. Production therefore varies considerably among the years. In the 2006/07 season for instance, rains were not timely, some areas were hit by hailstorms during crucial stages, and in other regions there was not enough moisture in the soil for boll formation. During that season, Uganda produced 75,000 metric tons[5] of seed cotton, recording an average yield of just 483 kilograms/hectare (FAO 2010). This production resulted in a total of 24,790 tons of lint (Cotton Development Organisation 2008). Even though the area cultivated increased from 100,000 hectares in 2005/06 to 150,000 hectares in 2006/07, yields declined in the same period. In the 2007/08 season there was a sharp decline in total cotton production. Total lint produced was only 12,303 tons, merely 65 percent of the total produced in the previous season.

---

5   All tons are metric tons in the chapter.

Although climatic events can considerably affect cotton performance, institutional factors are also crucial determinants of the low output and high variability. The productivity of cotton in Uganda is below international and regional averages, even though in the past years seed cotton yields have registered their highest records since 1960.[6] CDO acknowledges the limited availability of high-quality inputs, including seed, extension, and credit, as the major reason for the poor performance of the cotton subsector. Not only are inputs expensive or unaffordable for small producers, but their availability is also limited. Access to production inputs like fertilizers or high-quality seed is difficult to predict. Usually, the cotton season starts when seeds are made available to farmers by ginneries. Given the vertical integration and articulation of the chain, it is not uncommon that delays in the delivery of seed and other inputs occur. This situation was particularly severe for the 2007/08 season, when the ginneries dropped their production support program (Cotton Development Organisation 2008).

ORGANIC SYSTEM

Uganda and Tanzania are the first organic cotton producers in the African continent (Moseley and Gray 2008). In Uganda, organic cotton production started in about 1994 (Ogwang, Sekamatte, and Tindyebwa 2005) when the Export Promotion of Organic Product from Africa began the Lango Organic Project in Lira and Apac districts. The project was promoted by a Dutch trader in organic textiles (Bo Weevil BV), and it was established as a business-oriented enterprise (Tulip and Ton 2002). By 2004 there were almost 38,000 certified organic farmers in Uganda, almost one-third of them producing cotton (Taylor 2006).

In Lango, cotton is produced in rotation with sesame, an oilseed crop that commands a much higher productivity and market price compared to cotton. Farmers do not have many alternatives to cotton in this rotation system. In general, productivity of cotton in the northern region is considerably lower in the northernmost districts. In theory, organic ginneries buy the cotton from farmers at a premium price, but after discounting for the transport and ginning services, this premium price is not actually realized by the organic farmer. Dunavant is the only company that has certified organic ginneries. Other organic ginneries, namely Copcot and Lango Cooperative Union, segregate areas for organic production, according to their needs.

---

6 According to *FAOSTAT*, cotton seed yield in 2009 was greater than 900 kilograms/hectare (FAO 2010).

Table 3.1 shows the rapid expansion of organic cotton production since 2000. From 2000 to 2006, organic cotton production expanded at an average rate of 2 percent. This rate increased to 9 percent in the 2006/07 season. The main reasons for this positive trend were the growing interest of farmers in a crop system that requires lower use of often unaffordable inputs and the possibility of obtaining a premium price for their product. In the most recent season that we have information about (2007/08), the production share of organic cotton was even higher (20 percent), showing a still-growing interest in the organic system.

The main concern related to potential GM cotton introduction and adoption in Uganda is the coexistence of an organic system with a conventional system that is using GM seed. The positions of the government and of CDO toward organic production are not completely clear. In a way, the political support for this production system needs to be clarified before GM is introduced. Coexistence is possible, but the steps to support this coexistence have to be designed and implemented ahead of GM introduction to avoid contamination and thus damage to organic cotton exports.

Methodology

The evaluation of the potential impact of GM cotton seed in Uganda was done using partial budgets and stochastic analysis. Given that GM cotton has not yet been planted in Uganda, the study was based on assumptions about specific variables that determine cotton profitability, such as GM prices and adoption rates. These assumptions, summarized in Table 3.2, are crucial to the

TABLE 3.1   Organic cotton production, 2000–2008

Season	Seed cotton, organic (kg)	Seed cotton, national production (kg)	Lint, organic (bales)	Lint, national production (bales)	Share organic lint (%)
2000/2001	1,642,458	54,996,904	3,066	102,200	3.0
2001/02	1,734,187	63,898,025	3,406	126,148	2.7
2002/03	1,203,753	57,563,429	2,407	114,619	2.1
2003/04	2,030,465	84,344,870	3,626	160,000	2.3
2004/05	2,979,969	130,854,714	5,321	254,000	2.1
2005/06	1,499,030	51,847,138	2,677	102,000	2.6
2006/07	7,377,333	68,681,469	13,174	134,000	9.8
2007/08	n.a.	n.a.	13,766	66,500	20.7

Source: Cotton Development Organisation (2008).
Note: 1 bale = 185 kg. n.a. = not available.

TABLE 3.2   Assumptions for variables and distributions used for partial budget simulations

Component	Distribution	Assumptions and source
Yield (kg/ha)	@Risk Best-Fit distribution	Based on information collected from farmers.
Yield losses due to bollworms and lack of weeding (%)	@Risk Best-Fit distribution	Based on information collected from farmers.
Technology efficiency (%)	Triangular distribution	Values: low = 0, mean = 50, and high = 100, based on literature for both insect-resistant cotton and herbicide-tolerant cotton (Pray et al. 2002; Qaim 2003; Traxler and Godoy-Avila 2004).
Produce price (US$/kg)	@Risk Best-Fit distribution	Based on information collected from farmers.
Seed costs (US$/ha)	Not a distribution	Seed is distributed free of charge. The value assumed was UGX350/kg. On average, farmers use 4 kg/acre of seed for planting cotton.
Premium price (%)	Triangular distribution	For organic producers, percentage over official price: low = 0%, mode = 12.5%, and high = 25%.
Technology fee (%)	Triangular distribution	Scenario 1: Percentage over assumed price of formal seed: low = 0%, mode = 50%, and high = 100%. Scenario 2: Range of values found in the literature, including Falck-Zepeda, Traxler, and Nelson (2000); Huang et al. (2003, 2004); and Bennett et al. (2004). For comparison, 15 US$/ha corresponds to India, 32 US$/ha to South Africa and China, and 56 US$/ha to the United States.
Pesticide use (US$/ha)	@Risk Best-Fit distribution	Based on information collected from farmers.
Reduction in pesticide used to control Lepidoptera (%)	Triangular distribution	Values: low = 0%, mode = 50%, and high = 100%.
Herbicide use (US$/ha)	@Risk Best-Fit distribution	Based on information collected from farmers.
Increase in herbicide use (%)	Triangular distribution	Values: low = 0%, mode = 50%, and high = 100%. Increase over the average among current herbicide users.
Labor for pesticide application (US$/ha)	@Risk Best-Fit distribution	Based on information collected from farmers.
Reduction rate in labor used for pesticide application (%)	Triangular distribution	Values: low = 0%, mode = 25%, and high = 50%. The reduction in labor is related to the reduction in total pesticide applied.
Labor for herbicide application (US$/ha)	@Risk Best-Fit distribution	Based on information collected from farmers.
Increase rate in labor used for herbicide application (%)	Triangular distribution	Values: low = 0%, mode = 25%, and high = 50%. This value reflects an increase over the average among current herbicide users of the labor used to apply herbicides.

Source: Authors.

Notes: See tables in Appendixes A and B for details on the distribution of each variable. UGX = Ugandan shilling; US$ = US dollars.

results of the economic evaluation. The use of different scenarios based on the variability of these assumptions gives an idea of the scope of their impact. Few studies have explicitly recognized the year-to-year variability in farm profits by applying stochastic approaches (Pemsl, Waibel, and Orphal 2004; Hareau, Mills, and Norton 2006) that we address in this study.

Survey Instrument

A survey instrument was implemented to collect information on cotton production and current practices used. In addition to input use and production questions, the survey elicited information on subjective yield distributions from growers to gauge farmers' perceptions of the extent of yield losses caused by bollworms and by weeds. Farmers provided information on: (1) cotton yield without the constraint, (2) cotton yield with the constraint without using insecticides, and (3) cotton yield with the constraint and chemical control of the pest. We elicited low, mode, and high values from farmers. Each yield parameter (minimum, maximum, and mode) was used to assess variability both within and across observations.

The survey information enabled (1) calculation of partial budgets for representative growers and (2) simulation of partial budgets for various scenarios, including use of varieties that are genetically resistant to bollworm attacks and some other Lepidoptera pests and tolerant to herbicide applications.

Data Collection, Sites, and Sampling

Lira and Kasese are the districts where the GM confined trials have been implemented, and they are also the districts selected for evaluating the current cotton production systems and conducting our household interviews. After identifying cotton-producing districts, we randomly selected villages with farmers cropping cotton in the 2006 and 2007 seasons, three in Lira and seven in Kasese. The distribution of villages followed the proportion of cotton produced in the areas, but it also was intended to give a good representation of organic producers.

A total of 150 household heads were interviewed, 48 in Lira and 102 in Kasese. The households were randomly selected from the list of producers provided by ginneries operating in each region. This was the most complete list of cotton producers that we could access. Given the nature of the crop and the vertical integration of the system, all cotton producers need to sell their output to a ginnery. The questions were addressed for the 2007 campaign, and some additional recall information was collected for 2006. In some cases, selected producers cultivated more than one plot, but most of them only managed one plot of cotton. The information was analyzed per plot for the 2007 season only. Plots with incomplete information were not considered in the analysis. Thus, the total number of observations in our analysis ended up being 151 plots managed by 129 farmers. Of this total, 35 were plots from 32 producers located in Lira with only 12 real organic producers,[7] and the rest were plots from producers located in Kasese.

Stochastic Budget Analysis

We used basic partial budget analysis augmented with stochastic simulations to evaluate cotton profitability across different scenarios. The scenarios evaluated included (1) a conventional cotton producer, (2) an organic cotton producer, (3) a conventional producer using Bt cotton seed, (4) a conventional producer using HR cotton seed, and (5) a hypothetical case where an organic cotton producer is using a GM seed. Additionally, we artificially classified producers as "low input" and "high input" to get some insights about the effect of higher input use on cotton performance (see appendix tables 3A.1–3A.5 for details on the descriptive variables used).[8] Fertilizer use was considered as a criterion to classify high-input producers. Obviously, this classification is also a proxy for income level. Therefore, the category "high-input producers" refers to farmers who use chemical fertilizers and above-average amounts of pesticides in our sample. From a total number of 151 observations, only 27 qualified as high-input users.

The survey provided information to estimate input use and their costs and derive the partial budgets. The basis for calculating the partial budgets was the comprehensive guide produced by CIMMYT (1988). Cotton seed is distributed free of charge, and thus the value is zero for the producer. This information was used for the partial budgets of the conventional, organic, high-input, and low-input producers. For the simulated scenarios, we imputed seed costs based on the farmgate price of the cotton seed. Total use of chemical and organic fertilizers and pesticides was reported by farmers and converted to values per hectare. The value of the land was the equivalent of renting it. Average wages paid to hired labor were used to estimate the total family labor costs. This assumption seems reasonable in the production areas studied, where labor markets are active and farmers produce the crops commercially. Male and female labor days were valued equally. Using this information, we estimated

---

7 When analyzing the data, we noticed that several farmers who called themselves organic cotton producers were using insecticides and chemical fertilizers; therefore, the number of organic producers dropped considerably.

8 The use of statistical and econometric tests is a better alternative when there are a larger number of observations.

expected total income, total cost, expected net income, marginal benefit, and benefit/cost ratio for each of the five scenarios.

To introduce variability in the partial budgets, the study used @Risk software (Palisade Corporation, www.palisade.com/risk/) to estimate candidate distributions for each input variable. Note that in @Risk language, there are two kinds of variables: input variables, which are predetermined, and output variables, which are estimated based on the input variables. @Risk selected a best-fit distribution for input variables feasible to be obtained from farmers using the survey instrument. For input variables with limited or no information, we used triangular distributions, defined by low, mode, and high values. The triangular distribution is the simplest distribution to elicit from farmers, it approximates the normal distribution, and it is especially useful in cases where no sample data are available (Hardaker et al. 2004). For generation of the variable parameters (low, mode, and high values), we assumed values generated by expert consultation or literature review.

Input variables generated using survey information were yield, output price, pesticide use/cost, herbicide use/cost, and spraying cost (mainly labor). Input variables adjusted to triangular distributions were technology efficiency (trait expression), the technology fee, reduction rates in pesticide use, reduction rates in spraying costs in the case of Bt cotton, and increase rates in herbicide use for the case of HT cotton. Details on the minimum, mode, and maximum values adopted for these variables are reported in Table 3.2.

The @Risk software used the input variables to predict the distribution of the selected output variable (marginal benefit). In this way we not only compare marginal benefits across scenarios but also determine how sensitive the output variable is to changes in each input variable (within a scenario). A tornado graph is used to express the relative impact of a particular input parameter to the output from the simulations. The @Risk program regresses each output variable, in this case marginal benefits, to each of the parameters included in the simulation with a probability distribution. The resulting parameter gives an indication of the relative strength of the relationship between parameters and outcome.

Is Cotton Profitable?

Basic statistics of the household and production characteristics of interviewed farmers are presented in Table 3.3. In terms of cotton seed yield, aggregated values are higher than national averages for 2007 (FAO 2010), indicating some selection bias in the sampling.

TABLE 3.3 Descriptive statistics

Variable	Total sample (N = 151)		Lira (N = 35)		Kasese (N = 116)		t-test	P-value
	Mean	SE	Mean	SE	Mean	SE		
Gender of household head (female = 1)	0.09	0.02	0.03	0.03	0.11	0.03		
Control of plot (female = 1)	0.46	0.08	0.29	0.13	0.51	0.09		
Age of household head	44.04	1.14	42.63	2.85	44.47	1.22		
Education level of household head (years)	2.90	0.15	3.03	0.30	2.86	0.18		
Household size (number)	7.75	0.31	7.40	0.52	7.86	0.38		
Number of males older than 16	1.86	0.11	2.23	0.22	1.75	0.12		
Number of females older than 16	1.74	0.10	1.71	0.17	1.75	0.12		
Number of people younger than 16	4.15	0.23	3.46	0.33	4.36	0.28		
Experience with cotton (years)	14.68	1.04	16.86	2.36	14.02	1.15		
Probability of bollworm attacks	0.74	0.35	0.59	0.38	0.78	0.33	2.8926	0.004
Land value (US$)	1,192.90	2,330.54	1,167.10	1,634.88	1,200.68	2,508.77		
Total area (ha)	1.42	2.42	1.34	2.51	1.45	2.4		
Cotton area (ha)	0.68	0.55	0.44	0.24	0.75	0.59	3.0129	0.003
Seed cotton price (US$/kg)	0.39	0.06	0.39	0.08	0.39	0.05		
Seed cotton yield (kg/ha)	953.48	719.66	675.53	548.62	1,037.34	745.61	2.6592	0.009
Ouput value (US$/ha)	630.37	805.6	288.97	253.32	733.37	883.95	2.932	0.004
Dummy for organic producer			0.34	0.48				
Dummy for use of herbicides					0.09	0.29		
Herbicide use (US$/ha)					1.38	5.7		
Dummy for use of fertilizer	0.09	0.29	0.03	0.17	0.11	0.32		
Fertilizer use (US$/ha)	1.01	7.06	0.04	0.23	1.3	8.04		
Dummy for use of pesticides	0.97	0.16	0.89	0.32	1.0	0.0	3.8431	0.000
Pesticide use (US$/ha)	21.52	20.89	9.42	26.23	25.16	17.55	4.1079	0.000
Labor for weeding (US$/ha)	69.41	70.89	66.79	86.11	70.20	66.03		
Labor for herbicide application (US$/ha)					0.18	0.83		
Labor for pesticide application (US$/ha)	6.94	10.86	4.31	7.91	7.73	11.51		
Total hired labor (US$/ha)	147.52	127.98	164.62	179.92	142.35	108.07		

Source: Authors' calculations based on household survey information.

Notes: SE = standard error; US$ = US dollars. The t-test and P-values compare Lira and Kasese.

Sample statistics show that household characteristics are comparable across districts, but some production variables behave significantly differently. The age of the household head, level of education, household size, and household composition are similar across sites. Neither is there any significant variation concerning land value, labor use, and years of experience in cotton cultivation. The average farmer interviewed had more than 14 years' experience working on cotton.

The size of the cotton plot tends to be larger in Kasese. Similarly, seed cotton yield and total benefits generated from cotton production are also higher in Kasese. In contrast, it seems that bollworm infestation levels are higher in Kasese than in Lira, implying that Bt cotton could have a higher success in the Western than in the Northern region. In either case, chemical pesticides are not significantly reducing damage caused by pests or weeds (see Appendix C). A mean comparison of basic statistics between the two districts shows a significant difference in farmers' experiences in cotton production (Table 3.3).

Another person besides the household head can manage a cotton plot. So, although the percentage of female household heads in our sample is low (3 percent in Lira and 9 percent in Kasese), the share of plots managed by women can be as high as 50 percent in Kasese and 29 percent in Lira. Nevertheless, when tested for mean differences between plots managed by men or women, none of the variables analyzed behaved significantly differently (Table 3.4).

Traditional Production

The survey data analysis shows that most of the farmers are low-input users (Table 3.5). Some farmers use some type of chemical control to deal with insect pests, but relatively few make use of fertilizers, and almost none of them use herbicides. Cotton is a labor-intensive crop: labor represents more than 50 percent of the total production costs. Labor is primarily used for manual weeding. Weed infestation is therefore another serious constraint in cotton production. In our sample, weeding represents more than 20 percent of the total input costs for both types of producers. Other institutions working in cotton in the area report similar patterns (Agricultural Productivity Enhancement Program, pers. comm., 2008). Results do not confirm the general belief that women mainly handle weeding. In any case, cotton is a labor-consuming activity, and weeding is particularly labor demanding. So, freeing labor from weeding could allow family members to be available for other economic activities. However, the use of HT seed could have a

TABLE 3.4   Descriptive statistics of main variables, control of plot

Variable	Male-controlled plot (N = 112)		Female-controlled plot (N = 24)	
	Statistic	Standard error	Statistic	Standard error
Age of household head (years)	42.88	1.36	50.13	2.70
Education level of household head (years)	2.93	0.18	2.42	0.34
Land value (US$/ha)	2,660.02	471.26	2,782.70	514.64
Total area (ha)	1.38	0.21	1.35	0.20
Cotton area (ha)	1.62	0.11	1.35	0.15
Experience with cotton (years)	14.52	1.24	15.04	2.37
Probability of bollworm attacks	0.71	0.03	0.76	0.06
Seed cotton price (US$/kg)	0.38	0.00	0.38	0.00
Output value (US$/ha)	535.44	51.97	554.97	58.71
Seed cotton yield (kg/ha)	932.60	63.28	949.38	71.54
Labor used for weeding (US$/ha)	66.87	6.00	66.50	6.89
Total labor used (US$/ha)	140.54	10.91	140.85	12.61

Source: Authors' calculations based on household survey information.

Notes: We ran statistical tests, and none of these variables behave significantly differently between the two groups. Only 136 farmers reported on "who controls the plot," therefore the number of observations is lower than the total sample size. US$ = US dollars.

negative impact on employment and welfare in the community if there are not many off-farm labor opportunities. Almost all the farmers make use of hired labor.

Productivity of seed cotton for our sample, 386 kilograms/acre (or about 800 kilograms/hectare), is above the reported national average in 2007 (400 kilograms/hectare) but is lower than the average yield registered in 2009.[9]

On average, bollworms can reduce expected output by more than 70 percent. These estimations are based on farmers' perceptions and may have an upward bias, but they are a good reference to understand the severity of bollworm infestation in these regions. In addition to bollworms, there are other common biotic stresses, such as aphids, *Lygus* spp. (a sucking insect), and cotton stainers. These biotic constraints combined with high price variability

---

9   The *FAOSTAT* average for seed cotton for the past 5 years is about 678.92 kilograms/hectare and for the past 10 years is about 508.75 kilograms/hectare (FAO 2010). In 2009 the average yield was 942.8 kilograms/hectare.

TABLE 3.5   Cotton profitability for low- and high-input systems, season 2007/08

Cost component	Units	Low input (N = 124)	Share (%)	High input (N = 27)	Share (%)
Seed cotton, yield	kg/ha	918.46	—	1,132.78	—
Yield loss to bollworms	US$/kg	0.38	—	0.41	—
Total income	US$/ha	349.01	—	464.44	—
Land rent	US$/ha	75.63	30	69.16	19
Chemical fertilizer	US$/ha	—	—	24.56	7
Organic fertilizer	US$/ha	—	—	21.12	6
Herbicide use	US$/ha	—	—	17.16	5
Chemical pesticide	US$/ha	25.00	10	21.62	6
Labor to apply pesticides	US$/ha	6.88	3	5.88	2
Labor to apply herbicides	US$/ha	—	—	3.97	1
Labor for weeding	US$/ha	60.57	24	95.70	26
Labor for harvesting	US$/ha	25.93	10	34.97	10
Labor for other activities	US$/ha	54.46	22	72.98	20
Family labor	US$/ha	536.36	—	419.89	—
Total costs	US$/ha	248.47	—	367.13	—
Margin	US$/ha	100.54	—	97.31	—
Downside risk	%	43.60	—	41.30	—
Benefit/cost ratio		1.40	—	1.27	—

Source: Authors' calculations based on household survey information. The descriptive statistics based on type of producer are presented in Appendix D.

Notes: In both cases, when family labor is accounted for, the margins are negative and the downside risk is almost 100 percent. The benefit/cost ratio for low-input producers is 0.43, whereas that for high input producers is 0.59. US$ = US dollars.

and the unreliable availability of inputs make cotton production a very risky activity in Uganda. The estimated downside risk—the risk of not being able to cover at least the production cost—for surveyed farmers was more than 40 percent.

Looking at the main cost components, it is evident that farmers invest very little in fertilization. Most of the producers interviewed belong to the category "low-input producers," and although they do use pesticides to control for Lepidoptera and other major pests (*Lygus* spp., *Aphis,* and so forth), the amount of pesticide used is well below standard recommendations. In contrast, "high-input producers" reported not only higher yields but also higher prices paid for their cotton.

Organic versus Conventional Production

The main purpose for implementing the household survey in Lira was to cover a representative number of organic producers and generate a standard partial budget for an average organic cotton producer. However, only 12 of 35 producers interviewed complied with the standards of organic production. The rest of the producers admitted using some level of chemical control to deal with heavy pest infestations. In fact, the number of organic farmers in Uganda changes from year to year, as farmers appear to switch from conventional to organic with relative freedom. According to Dunavant representatives, in the 2006/07 season, 11,691 organic farmers were registered and contracted for a total production of 6,600 bales, which accounted for almost one-third of the total production of this company. The number of organic producers, however, dropped significantly during the 2008/09 season (see Table 3.1), as there were serious problems with army bollworms that infested the organic fields. During field observations, it became clear that organic farmers were so concerned about pest attacks that many applied pesticides even if they were not supposed to. Statistical analysis over this small sample of producers would obviously not render meaningful results, but the analysis of the household surveys can still provide some useful insights.

As for any other organic crop, cotton requires a significant amount of labor for manual activities, including insect and weed control. In our sample, labor use represented more than 60 percent of the total cost of production for organic cotton producers, and about one-third of that was labor used in manual weeding. Surveyed farmers reported that the damage caused by bollworms was more than 50 percent, which is comparatively lower than the reported 76 percent for conventional producers. This difference could be because organic producers do manual control of bollworms, an activity that is quite labor demanding but more effective than the sporadic application of pesticides. Approximately 12 percent of the total investment of conventional producers was on chemical pesticides, whereas organic producers only spent 2 percent on organic pesticides. Organic pesticides include the use of neem extract and other nonchemical applications.

Given these management practices, conventional producers recorded 17 percent margins, whereas organic producers had 5 percent. It is well known that the productivity of organic cotton is lower than that of conventional production (Ogwang, Sekamatte, and Tindyebwa 2005). This low productivity is somehow compensated for by a price premium that the organic producers get for their output. Yet the organic cotton farmers interviewed were not getting these price premiums, as the prices reported were comparable to those received by conventional producers. Not only was the profitability of cotton low for both types of producers, but the crop also showed a high downside risk. In the case of organic farming, the downside risk for the 12 farmers interviewed was higher than 50 percent, and for conventional producers it was about 39 percent (Table 3.6).

TABLE 3.6   Cotton profitability for conventional and organic cotton producers, 2007/08 season

Cost component	Units	Scenario 1: Conventional (*N* = 139)	Share (%)	Scenario 2: Organic (*N* = 12)	Share (%)
Yield	kg/ha	962.1	—	863.5	—
Price reported by farmers	US$/kg	0.38	—	0.38	—
Total income	US$/ha	367	—	328	—
Land rent	US$/ha	74.99	22	72.25	26
Chemical fertilizer	US$/ha	24.56	7	0.00	—
Organic fertilizer	US$/ha	20.23	6	22.01	8
Herbicide use	US$/ha	17.16	5	0.00	—
Pesticide to control Lepidoptera	US$/ha	23.61	7	—	—
Pesticide to control other pests	US$/ha	17.44	5	—	—
Organic pesticide	US$/ha	—	—	4.82	2
Hired labor to apply pesticides	US$/ha	7.24	2	3.37	1
Hired labor to apply herbicides	US$/ha	5.42	2	0.00	—
Hired labor for weeding	US$/ha	70.91	20	61.20	22
Hired labor for harvesting	US$/ha	28.92	8	28.90	10
Hired labor for other activities	US$/ha	57.49	17	89.09	32
Family labor	US$/ha	423.98	—	1,652.39	—
Total costs	US$/ha	347.98	—	281.66	—
Margin	US$/ha	18.78	—	46.56	—
Downside risk	%	38.6	—	52.0	—
Benefit/cost ratio		1.05	—	1.17	—

Source: Authors' calculations based on household survey information.
Notes: When family labor is accounted for, the margins are negative and the downside risk is almost 100 percent. The benefit/cost ratio for conventional producers is 0.48, whereas that for organic producers is 0.17.
— = not applicable; US$ = US dollars.

GM Cotton

The use of GM seed was simulated in three scenarios (Table 3.7): Bt cotton, HT cotton, and organic cotton plus Bt cotton. Despite the seed-technology advantage, the profitability of cotton production was still very low for all these scenarios but higher than the situation with no technological change.

The use of GM seed may reduce the downside risk, but this depends on the effectiveness of the technology to control the constraint (e.g., the expression of the trait). An expert from the CDO (D. Lubwana, agronomist with CDO, pers. comm., 2007) reported that yield losses due to bollworms can be as high as 80 percent, which is in agreement with what farmers reported in the survey (on average, about 76 percent). Given these high values, and farmers' perceptions of yield losses due to bollworm attacks and weed infestation, it is not surprising that the margins are higher for both the Bt and the HT cotton scenarios. Perceptions, however, are usually biased upward, given that it is rather difficult for farmers to isolate the effect of one constraint. The marginal benefits of using GM seed are directly related to the level of incidence of the productivity constraint and the actual damage caused by the biotic constraint.

The HT scenario recorded the highest benefit/cost ratio and the highest marginal rate of return of all the likely scenarios.[10] The explanation for this result lies in the assumptions used for simulating this scenario. As mentioned earlier, weeding is a labor-intensive activity that needs to be performed regularly in a cotton field. Failure to weed has a severe impact on the final yields. Unfortunately, the lack of technical information about weed infestation and weed control in cotton production in Uganda is a major constraint on the model. Thus, the assumptions for this scenario are based on expectations rather than on technical information. In comparison, the impact of Bt cotton has been more thoroughly documented, and there is more information to support the assumptions behind this scenario.

It is important to point out here that a higher yield would probably demand higher labor for agricultural practices, mainly for harvesting. Thus, the margins of all the simulated scenarios are likely to be overestimated. We do not have enough technical information to support more specific assumptions about the potential increase in labor costs, as there can be many factors affecting labor use. For instance, if we assume that the labor used for harvesting is proportional to yield increases, we may be overlooking the fact that labor may

---

10 Given the controversy between organic system proponents and GM technology supporters, the organic plus GM seed scenario is quite unlikely to be implemented.

not be readily available at the critical times or that the household may not have enough financial resources to hire more labor. As shown in Table 3.7, yield increases imputed to GM seed are basically due to the combination of damage-control effects and the technology efficiency, which could be used to represent the upper bound of additional labor used for harvesting under perfect correlation. Taking the average values shown in Table 3.7, Bt cotton reports the highest reduction in marginal benefits (about 30 percent) whereas the organic plus Bt scenario reports the lowest reduction (13 percent). In other words, if the producer uses mainly hired labor, the practice will definitely have an impact on the margins. This type of producer will also invest in other complementary inputs that will contribute to an even better performance of the variety, and thus compensate for the additional investment.

Figure 3.3 presents a graphical analysis of the marginal benefits for all five scenarios evaluated. The distribution of marginal benefits is represented in the histogram, and the tornado graph summarizes the relative impact of a particular input variable on these margins.[11] For all scenarios the variability in yield and the high labor costs are the main determinants of the margins generated. A technology that contributes to reduce this yield variability would definitely have an impact on farmers' welfare.

Conclusions and Policy Recommendations

This study provides an ex ante evaluation of the potential impact of GM cotton adoption in Uganda. A survey was used to calculate partial budgets for representative growers and compare partial budgets for various real and simulated scenarios. The partial budget of a low-input cotton producer was compared with that of a high-input producer. Similarly, we compared the partial budget of a conventional cotton producer with that of an organic producer. The latter two cases were used to develop the simulated scenarios of conventional cotton producers using GM seed (both Bt and HT cotton) and organic producers using Bt cotton.

The primary data for this analysis comes from two main cotton-producing districts in Uganda: Kasese and Lira. Partial budgets are used to evaluate the profitability of cotton production and to compare conventional and organic cotton production with hypothetical GM scenarios. We added stochastic simulations to the partial budgets to account for the effects of risk

---

11 Appendix E presents the graphical analysis of the marginal benefits for the low- and high-input systems.

TABLE 3.7   Partial budgets for scenarios using genetically modified seed

Cost component	Units	Insect-resistant cotton	Herbicide-tolerant cotton	Organic + premium price	Organic + Bt
Yield	kg/ha	1,325.40	1,342.60	863.5	1,101.34
Yield loss to bollworm attacks	%	76	—	55	55
Yield loss to weeds	%	—	79	—	—
Technology efficacy	%	50	50	—	50
Price reported by farmers	US$/kg	0.38	0.38	0.43	0.43
Premium price	%	—	—	12.5	12.5
Total income	US$/ha	505.27	511.83	369.24	470.97
Seed cost (4 kg/acre)	US$/ha	1.58	1.58	0.00	1.58
Land rent	US$/ha	74.99	74.99	72.25	72.25
Chemical fertilizer	US$/ha	24.56	24.56	0.00	0.00
Organic fertilizer	US$/ha	20.23	20.23	22.01	22.01
Herbicide use	US$/ha	17.16	25.74	0.00	0.00
Increase rate of herbicide use	%	—	50	—	—
Pesticide to control Lepidoptera	US$/ha	11.81	23.61	—	—
Reduction rate in pesticide use	%	50	—	—	—
Pesticide to control other pests	US$/ha	17.44	17.44	—	—
Chemical pesticide	US$/ha	—	—	0.00	0.00
Organic pesticide	US$/ha	—	—	4.82	4.82
Labor to apply pesticides	US$/ha	5.43	7.24	3.37	2.53
Reduction rate in labor costs	%	25	—	—	25
Labor to apply herbicides	US$/ha	5.42	8.13	0.00	0.00
Increase rate in labor costs	%	—	50	—	—
Labor for weeding	US$/ha	70.91	35.45	61.20	61.20
Reduction rate in labor costs	%	—	50	—	—
Labor for harvesting	US$/ha	28.92	28.92	28.90	28.90
Labor for other activities	US$/ha	57.49	57.49	89.09	89.09
Family labor	US$/ha	423.98	423.98	1,652.39	826.19
Total costs	US$/ha	335.94	325.40	281.66	282.39
Margin	US$/ha	169.33	186.44	87.58	188.57
Downside risk	%	26.8	21.6	48.4	40.8
Benefit/cost ratio		1.50	1.57	1.31	1.67

Source: Authors' calculations based on household survey information.

Notes: In all cases, when family labor is accounted for, the margins are negative and the downside risk is almost 100 percent. The benefit/cost ratio for the insect-resistant cotton scenario is 0.66, whereas that for the herbicide-tolerant cotton scenarios is 0.68, and for the organic scenario plus insect-resistant cotton is 0.42. Except for expected changes in pesticides/herbicides and labor used to apply them, the information and distributions are the same as those for the case of conventional producers (see Table 3.6). Except for expected changes in (organic) pesticides and labor used to apply them, the information and distributions are the same as those for the case of organic producers (see Table 3.6).
— = not applicable; Bt = insect resistant; US$ = US dollars.

FIGURE 3.3 Graphical analysis of marginal benefits. (a) low-input producer; (b) high-input producer; (c) conventional producer

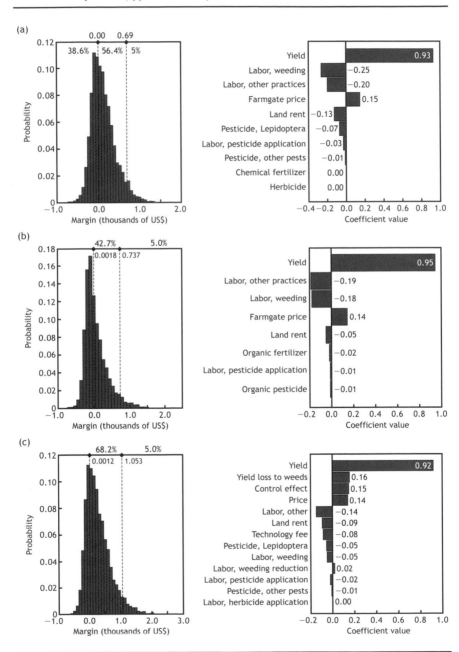

FIGURE 3.3   Continued. Graphical analysis of marginal benefits. (d) organic producer; (e) insect-resistant cotton producer

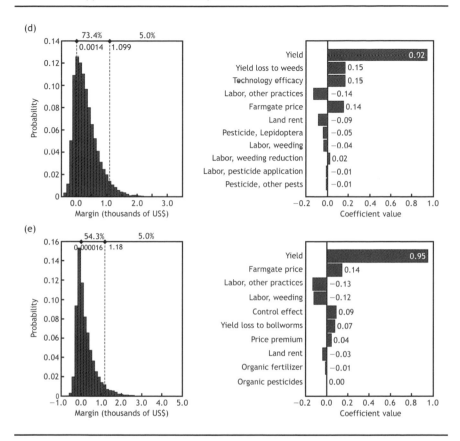

and uncertainty related to cotton production and profitability. The sampling framework for the collection of the primary information does not permit the extension of findings to the country level. However, this exercise does provide an initial assessment of the potential impact of GM cotton adoption on farmers in the most important cotton districts in Uganda.

Our findings suggest that GM cotton has the potential to contribute to improving the productivity of cotton. However, this technology would probably make only a minor contribution to the performance of cotton in Uganda if the access to high-quality inputs was not properly addressed. Overall, the simulations show that conventional production using Bt and HT cotton varieties could yield higher returns than either the conventional or the organic system, but the profitability of the crop will not increase dramatically. Our results also show that the largest benefits could be reached in the unlikely case that GM seed (Bt cotton) could be used in organic production. We have mentioned several times that access to high-quality inputs is critical to improve cotton performance in Uganda. From the perspective of a conventional production, this alternative deserves consideration. On the one hand, the current conventional cotton production uses minimal amounts of inputs and approximates the organic system. On the other hand, the investment and institutional changes needed to reach this goal are probably not feasible in the short run. Although our sample size, especially the number of observations for organic producers, does not allow us to make extensive conclusions, this is an area that needs further exploration.

The vertical integration of the chain could facilitate the dissemination of GM technology, but the availability of seed and inputs of good quality and appropriate extension support have to be guaranteed. Some problems exist with respect to seed quality, and it is not clear how the introduction of GM seed could solve these problems. For instance, it is necessary to improve the ginning quality in the country. Regardless of the type of seed or farming system used, investment in fertilizers, high-quality seed, and other agronomic and management practices are crucial to improve the profitability of cotton in Uganda.

In the case of Bt and HT cotton, farmers are not using significant levels of pesticides and herbicides, and therefore the expected reduction of both chemicals would be small. If yield losses due to bollworms are lower than reported by farmers (as in a low-incidence year), then the profitability of this technology will dramatically decrease. In this regard, it does make sense to frame the introduction of the GM cotton technology as an insurance protecting Ugandan farmers from catastrophic or severe losses stemming from target pests and weeds.

The assessment of other sectors of the economy, national industry, trading status, and the institutional environment also need to be addressed to get a full picture of the potential impact of GM cotton in Uganda. For instance, at the national level, the technology fee charged by the innovator largely affects the benefits generated by the adoption of GM cotton. Because cotton seed is currently distributed free of charge, it is clear that, before releasing the technology, Uganda needs to develop a strategy to negotiate with the developer of the technology the fee that will be charged to farmers. Similarly, the coexistence of organic and conventional cotton production using GM seed could be possible if institutional arrangements are implemented beforehand to avoid contamination and damaging organic exports.

A critical point in the case of the HT cotton scenario is farmers' abilities to apply the herbicide glyphosate. Very few farmers in the survey reported herbicide use, much less glyphosate, either because of the high costs of this input or its unavailability. If farmers cannot access glyphosate, then they will probably revert to manual weeding, with a direct impact on yield and expected benefits. There is also the issue of training in the use of herbicides and understanding the concept of herbicide tolerance and its implications for herbicide use. This raises a policy question that developers and policymakers in Uganda must address to ensure capturing the potential benefits of the technology. It would be worthwhile to explore in more detail whether a program to improve training and access to herbicides is warranted.

A question partially addressed by this study is whether reducing downside risk to cotton farmers in Uganda by reducing insect and weed damage compensates for the cost and effort in introducing a complex technology. It is important to take into account the additional inputs needed to enable damage control (such as the herbicide glyphosate), and changes in management practices (such as the need for scouting to ensure that insect pressures do not overwhelm Bt protection level) will determine how transferable the GM technology is and how easily it can be adopted by Ugandan farmers.

Although it is possible to compare the profitability of a given year of organic cotton production with conventional cotton production using GM seed, this gives only a small part of the picture. Because we are interested in contributing to poverty alleviation, it is much more significant—but at the same time challenging—to evaluate the long-term contribution of either system to the farmers' welfare. This is a research topic that needs further attention.

APPENDIX 3A: Details of Distributions of the Variables in the Partial Budgets of Real Scenarios

In all tables, the distributions are best-fit distributions using the @Risk software.

TABLE 3A.1   All sample (*N* = 151)

Variable name	Minimum	Mean	Maximum	Confidence interval 5%	95%
Yield (kg/ha)	1.4	951.7	6,406.7	129.8	2,401.6
Yield loss due to bollworm (%)	0.9	66.7	100	22.3	97.5
Price (US$/kg)	0.2	0.4	0.5	0.3	0.4
Land value (US$/ha)	0.0	730.3	8,502.7	37.4	2,186.6
Land rent (US$/ha)	0.0	33.9	316.2	1.7	101.6
Chemical fertilizer (US$/ha)	0.0	1.0	9.3	0.1	3.0
Organic fertilizer (US$/ha)	0.0	0.6	5.3	0.0	1.7
Herbicide (US$/ha)	0.0	18.9	37.8	1.9	35.9
Chemical pesticide (US$/ha)	0.0	20.4	240.7	1.0	61.2
Organic pesticide (US$/ha)	0.0	0.6	5.5	0.0	1.7
Labor used to apply pesticide (US$/ha)	0.0	6.8	72.3	0.3	20.3
Labor used to apply herbicide (US$/ha)	0.0	0.1	1.3	0.0	0.4
Labor used for weeding (US$/ha)	0.0	65.5	4623.0	3.4	196.1
Labor used for other operations (US$/ha)	0.5	74.8	483.2	12.8	178.8

Source: Authors.
Note: US$ = US dollars.

TABLE 3A.2   Low-input producer (N = 124)

Variable name	Minimum	Mean	Maximum	Confidence interval	
				5%	95%
Yield (kg/ha)	0.1	915.3	9,156.9	46.9	2,741.1
Yield loss due to bollworm (%)	0.6	66.7	100.0	22.4	97.5
Price (US$/kg)	0.2	0.4	0.6	0.3	0.5
Land value (US$/ha)	0.0	706.4	7,517.3	36.2	2,115.0
Land rent (US$/ha)	0.0	36.9	374.6	1.9	110.5
Chemical pesticide (US$/ha)	0.0	20.8	223.4	1.1	62.2
Labor used to apply pesticide (US$/ha)	0.0	6.9	64.4	0.4	20.6
Labor used for weeding (US$/ha)	0.0	60.8	580.2	3.1	182.0
Labor used for other operations (US$/ha)	0.0	54.5	518.9	2.8	163.1

Source: Authors.
Note: US$ = US dollars.

TABLE 3A.3   High-input producer (N = 27)

Variable name	Minimum	Mean	Maximum	Confidence interval	
				5%	95%
Yield (kg/ha)	0.0	1,133.0	13,414.2	58.1	3,392.7
Yield loss due to bollworm (%)	34.4	78.3	100.0	55.4	100.0
Price (US$/kg)	0.2	0.4	0.5	0.4	0.5
Land value (US$/ha)	0.0	828.8	8338.0	42.5	2,482.4
Land rent (US$/ha)	0.0	17.9	178.7	0.9	53.7
Chemical fertilizer (US$/ha)	0.0	5.5	55.0	0.3	16.3
Organic fertilizer (US$/ha)	0.0	3.1	31.9	0.2	9.4
Herbicide (US$/ha)	0.0	5.7	66.5	0.3	17.1
Chemical pesticide (US$/ha)	0.0	17.6	175.8	0.9	52.8
Labor used to apply pesticide (US$/ha)	0.0	8.4	24.8	0.6	19.4
Labor used to apply herbicide (US$/ha)	0.0	0.7	9.8	0.0	2.2
Labor used for weeding (US$/ha)	0.0	95.7	1,028.2	4.9	286.5
Labor used for other operations (US$/ha)	0.0	73.0	755.3	3.7	218.6

Source: Authors.
Note: US$ = US dollars.

TABLE 3A.4   Conventional producer (*N* = 139)

Variable name	Minimum	Mean	Maximum	Confidence interval 5%	95%
Yield	0.6	962.1	5,293.3	123.1	2,327.9
Yield loss caused by bollworm (%)	9.6	75.3	100.0	43.2	100.0
Yield loss caused by weeds (%)	7.0	79.1	100.0	31.8	100.0
Price (US$/kg)	0.1	0.4	0.5	0.3	0.4
Land value (US$/ha)	7.5	684.7	32,868.4	30.8	2,950.8
Land rent (US$/ha)	0.0	35.9	358.1	1.8	107.4
Chemical fertilizer (US$/ha)	0.0	1.1	10.6	0.1	3.2
Organic fertilizer (US$/ha)	0.0	0.3	2.7	0.0	0.9
Herbicide (US$/ha)	0.0	1.1	10.9	0.1	3.3
Pesticide to control Lepidoptera (US$/ha)	0.0	19.7	218.8	1.0	59.0
Pesticide to control other pests (US$/ha)	0.0	2.8	29.7	0.1	8.3
Labor used to apply pesticide (US$/ha)	0.0	7.2	90.4	0.4	21.7
Labor used to apply herbicide (US$/ha)	0.0	0.2	2.2	0.0	0.7
Labor used for weeding (US$/ha)	0.0	70.9	730.8	3.6	212.4
Labor used for other operations (US$/ha)	0.1	74.7	580.1	10.9	185.5

Source: Authors.
Note: US$ = US dollars.

TABLE 3A.5   Organic producer (*N* = 12)

Variable name	Minimum	Mean	Maximum	Confidence interval 5%	95%
Yield (kg/ha)	0.0	863.4	8,840.0	44.2	2,586.7
Yield loss due to bollworm (%)	0.1	59.3	88.9	19.9	86.6
Price (US$/kg)	0.2	0.4	0.7	0.3	0.5
Land value (US$/ha)	0.0	1,228.3	12,561.9	62.9	3,679.1
Land rent (US$/ha)	0.0	24.1	71.7	1.8	56.1
Organic fertilizer (US$/ha)	0.0	8.7	25.9	0.7	20.2
Organic pesticide (US$/ha)	0.0	2.3	6.8	0.2	5.3
Labor used to apply pesticide (US$/ha)	0.0	3.4	31.3	0.2	10.1
Labor used for weeding (US$/ha)	0.0	61.2	588.5	3.1	183.3
Labor used for other operations (US$/ha)	3.4	118.8	595.4	35.8	240.9

Source: Authors.
Note: US$ = US dollars.

APPENDIX 3B: Details of Distributions of the Variables in the Partial Budgets of Simulated Scenarios (%)

Scenario	Minimum	Mean	Maximum	Confidence interval	
				5%	95%
Control effect	1	50	100	16	84
Technology fee	1	50	100	16	84
Pesticide use reduction	0	50	100	16	84
Labor used to apply pesticide, reduction	0	25	50	8	42
Labor used to apply herbicide, increase	1	50	99	16	84
Labor used for weeding, reduction	50	75	100	58	92
Price premium	0	13	25	4	21
Labor cost reduction	0	25	50	8	42

Source: Authors.

Note: The distributions are best-fit distributions using the @Risk software.

APPENDIX 3C: Production Function Using a Damage-Abatement Specification

Variable name	Coefficient	Standard error	*t*-value
Yield			
Production function			
Constant	198.56	37.72	5.26 ***
District (dummy, Kasese = 1)	219.46	83.22	2.64 ***
Altitude (meters above sea level)	−0.25	0.09	−2.76 ***
Organic producer (dummy)	242.50	171.51	1.41
Land rent (US$/ha)	−0.61	1.37	−0.45
Square of land rent	0.00	0.00	0.73
Family labor (US$/ha)	0.38	0.16	2.41 **
Square of family labor	0.00	0.00	−2.07 **
Fertilizer (US$/ha)	11.61	9.56	1.21
Square of fertilizer	0.00	0.00	−0.30
Hired labor for harvesting (US$/ha)	2.19	1.66	1.32
Square of hired labor for harvesting	0.00	0.00	−0.16
Hired labor for other activities (US$/ha)	4.72	0.83	5.68 ***
Square of hired labor for other activities	0.00	0.00	−4.32 ***
Damage abatement			
Constant	10.54	10.57	1.00
Pesticide and labor used to apply pesticides (US$/ha)	0.57	0.57	1.01
Herbicide and labor used in weeding (US$/ha)	0.07	0.07	0.95

Source: Authors' calculations based on household survey information.

Notes: * denotes significance at the 10 percent level, ** at the 5 percent level, and *** at the 1 percent level. R2 = 0.45; adjusted R2 = 0.38. US$ = US dollars.

APPENDIX 3D: Descriptive Statistics of the Main Variables, by Type of Producer

Variable name	Low-input producers ($N = 124$)		High-input producers ($N = 27$)		F	Significance
	Statistic	Standard error	Statistic	Standard error		
Gender of household head (female = 1)	0.08	0.27	0.15	0.36		
Age of household head (years)	43.49	13.39	46.22	17.33		
Education level of household head (years)	2.77	1.86	3.44	1.89		
Land value (US$/ha)	2,568.01	5,708.80	3,929.03	4,523.43		
Total area (ha)	1.30	2.31	1.90	2.80		
Cotton area (ha)	0.69	0.58	0.64	0.30		
Experience with cotton (years)	13.73	12.30	18.81	14.30	3.6	*
Probability of bollworm attacks	0.74	0.35	0.67	0.36		
Seed cotton price (US$/kg)	0.38	0.04	0.41	0.03	31.3	***
Output value (US$/ha)	583.81	808.54	713.27	580.04		
Seed cotton yield (kg/ha)	918.46	715.10	1,132.78	705.57		
Labor used for weeding (US$/ha)	60.57	65.07	95.70	75.28	3.6	**
Total labor used (US$/ha)	132.45	112.67	188.25	157.35	3.0	**

Source: Authors' calculations based on household survey information.

Notes: * denotes significance at the 10 percent level, ** at the 5 percent level, and *** at the 1 percent level. US$ = US dollars.

APPENDIX 3E: Graphical Analysis of Marginal Benefits
Source: Authors.

Low-Input Producer

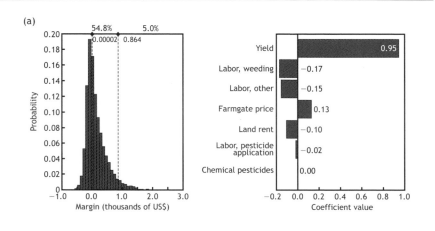

Source: Authors.

High-Input Producer

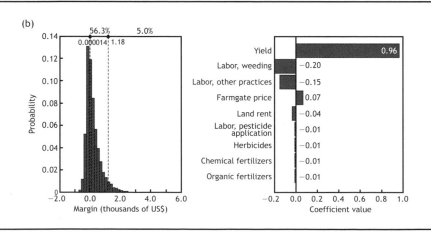

Source: Authors.

References

ACE (Audit and Control Experience). 2006. "End of Cotton Season Report 2005/06." Submitted to Uganda Ginners and Cotton Exporters Association, Kampala.

Baffes, J. 2009. "The 'Full Potential' of Uganda's Cotton Industry." *Development Policy Review* 27: 67–85.

Bennett, R., Y. Ismael, U. Kambhampati, and S. Morse. 2004. "Economic Impact of Genetically Modified Cotton in India." *AgBioForum* 7 (3): 96–100.

Cabanilla, L. S., T. Abdoulaye, and J. H. Sanders. 2005. "Economic Cost of Non-adoption of Bt Cotton in West Africa: With Special Reference to Mali." *International Journal of Biotechnology* 7: 46–61.

CIMMYT (International Maize and Wheat Improvement Center). 1988. *From Agronomic Data to Farmer Recommendations: An Economics Training Manual.* Mexico, D.F.

Cotton Development Organisation. 2006. *The Cotton Sector in Uganda: Progess Made in the Sector and Recommendations for Achieving Further Progress.* Kampala.

———. 2008. *Cotton Development Organisation Annual Report 2007/08.* Kampala.

De Groote, H., B. Overholt, L. J. Ouma, and S. Mugo. 2003. "Assessing the Potential Impact of Bt Maize in Kenya Using a GIS-Based Model." Paper presented at the 25th International Conference of the International Association of Agricultural Economists, August 16–22, in Durban, South Africa.

Edmeades, S., and M. Smale. 2006. "A Trait-Based Model of the Potential Demand for a Genetically Engineered Food Crop in a Developing Economy." *Agricultural Economics* 35: 351–361.

Elbehri, A., and S. MacDonald. 2004. "Estimating the Impact of Transgenic Bt-Cotton on West and Central Africa: A General Equilibrium Approach." *World Development (Oxford)* 32: 2049–2064.

Falck-Zepeda, J., J. D. Horna, and M. Smale. 2008. "Betting on Cotton: Potential Payoffs and Economic Risks of Adopting Transgenic Cotton in West Africa." *African Journal of Agricultural and Resource Economics* 2: 188–207.

Falck-Zepeda, J., G. Traxler, and R. G. Nelson. 2000. "Surplus Distribution from the Introduction of a Biotechnology Innovation." *American Journal of Agricultural Economics* 82 (2): 360–369.

Falck-Zepeda, J., J. D. Horna, P. Zambrano, and M. Smale. 2008. "Policy and Institutional Factors and the Distribution of Economic Benefits and Risk from the Adoption of Insect Resistant (Bt) Cotton in West Africa." *Asian Biotechnology Development Review* 11: 1–32.

FAO (Food and Agriculture Organization of the United Nations). 2010. *FAOSTAT. Production: Crops.* Accessed May 31. http://faostat.fao.org/. 2010.

Gordon, A., and A. Goodland. 2000. "Production Credit for African Smallholders: Conditions for Private Provision." *Savings and Development* 24 (1): 55–84.

Hardaker, J. B., R.B.M. Huirne, J. R. Anderson, and G. Lien. 2004. *Coping with Risk in Agriculture.* Wallingford, UK: CAB International.

Hareau, G. G., B. F. Mills, and G. W. Norton. 2006. "The Potential Benefits of Herbicide Resistant Transgenic Rice in Uruguay: Lessons for Small Developing Countries." *Food Policy* 31: 162–179.

Horna, J. D., M. Smale, R. Al-Hassan, J. Falck-Zepeda, and S. E. Timpo. 2008. *Insecticide Use on Vegetables in Ghana: Would GM Seed Benefit Farmers?* IFPRI Discussion Paper 00785. Washington, DC.

Huang, J., R. Hu, C. Pray, F. Qiao, and S. Rozelle. 2003. "Biotechnology as an Alternative to Chemical Pesticides: A Case Study of Bt Cotton in China." *Agricultural Economics* 29 (1): 55–67.

Huang, J., R. Hu, H. van Meijl, and F. van Tongeren. 2004. "Biotechnology Boost to Crop Productivity in China: Trade and Welfare Implications." *Journal of Development Economics* 75 (1): 27–54.

ICAC (International Cotton Advisory Committee). 2008. *World Cotton Trade.* Washington, DC.

James, C. 2011. *Global Status of Commercialized Biotech/GM Crops.* Ithaca, NY, US: International Service for the Acquisition of Agri-Biotech Applications.

Langyintuo, A. S., and J. Lowenberg-DeBoer. 2006. "Potential Regional Trade Implications of Adopting Bt Cowpea in West and Central Africa." *AgBioForum* 9 (2): 111–120.

Moseley, W. G., and L. C. Gray. 2008. *Hanging by a Thread: Cotton, Globalization and Poverty in Africa.* Athens, OH, US: Ohio University Press.

Ogwang, J., M. B. Sekamatte, and Tindyebwa. 2005. "Report on the Ground Situation of Organic Cotton Production in Selected Areas of the Lango Sub-region." National Agricultural Research Organisation, Kampala.

Pemsl, D., H. Waibel, and J. Orphal. 2004. "A Methodology to Assess the Profitability of Bt-Cotton; Case Study Results from the State of Karnakata, India." *Crop Protection* 23: 1249–1257.

Pray, C., J. Huang, R. Hu, and S. Rozelle. 2002. "Five Years of Bt Cotton in China—The Benefits Continue." *Plant Journal* 31: 423–430.

Qaim, M. 2003. "Bt Cotton in India: Field Trial Results and Economic Projections." *World Development* 31: 2115–2127.

Serunjogi, L. K., P. Elobu, G. Epieru, V.A.O. Okoth, M. B. Sekamatte, J. P. Takan, and J.O.E. Oryokot. 2001. "Traditional Cash Crops: Cotton (*Gossipium* sp.)." In *Agriculture in Uganda: Crops,* edited by J. K. Mukiibi, 322–375. Kampala, Uganda: Fountain Publishers / Technical Centre for Agriculture and Rural Cooperation (CTA) / National Agricultural Research Organisation.

Smale, M., P. Zambrano, G. Gruère, J. Falck-Zepeda, I. Matuschke, J. D. Horna, L. Nagarajan, et al. 2009. *Measuring the Economic Impacts of Transgenic Crops in Developing Agriculture during the First Decade: Approaches, Findings, and Future Directions.* Washington, DC: International Food Policy Research Institute.

Taylor, A. 2006. *Overview of the Current State of Organic Agriculture in Kenya, Uganda and the United Republic of Tanzania and the Opportunities for Regional Harmonization.* New York and Geneva: United Nations.

Traxler, G., and S. Godoy-Avila. 2004. "Transgenic Cotton in Mexico." *AgBioForum* 7 (1–2): 57–62.

Tschirley, D., C. Poulton, and P. Labaste. 2009. *Organization and Performance of Cotton Sectors in Africa.* Washington, DC: World Bank.

Tulip, A., and P. Ton. 2002. *Organic Cotton Uganda Case Study: A Report for PAN UK's Pesticides and Poverty Project.* London: Pesticide Action Network.

UEPB (Uganda Export Promotion Board). 2007. *Export Performance Watch.* Export Bulletin Edition 10. Kampala.

Uganda. 2007. "Country Statement." Presented at the 66th plenary meeting of the International Cotton Advisory Committee, October 19–26, Izmir, Turkey.

Vitale, J., H. Glick, J. Greenplate, and O. Traore. 2008. "The Economic Impacts of Second Generation Bt Cotton in West Africa: Empirical Evidence from Burkina Faso." *International Journal of Biotechnology* 10: 167–183.

Vitale, J. D., G. Vognan, M. Ouattarra, and O. Traore. 2010. "The Commercial Application of GMO Crops in Africa: Burkina Faso's Decade of Experience with Bt Cotton." *AgBioForum* 13 (4): 320–332.

You, L., and J. Chamberlin. 2004. *Spatial Analysis of Sustainable Livelihood Enterprises of Uganda Cotton Production.* Environment and Production Technology Division Discussion Paper 121. Washington, DC: International Food Policy Research Institute.

Chapter 4

Benefits, Costs, and Consumer Perceptions of the Potential Introduction of a Fungus-Resistant Banana in Uganda and Policy Implications

Enoch M. Kikulwe, Ekin Birol, Justus Wesseler, and José Falck-Zepeda

Banana is a staple crop in Uganda. Ugandans have the highest per capita consumption of cooking bananas in the world (Clarke 2003). However, banana production in Uganda is limited by several productivity constraints, such as insects, diseases, soil depletion, and poor agronomic practices. To address these constraints, the country has invested significant resources in research and development (R&D) and other publicly funded programs, pursuing approaches over both the short and long term. Uganda formally initiated its short-term approach in the early 1990s; it involves the collection of both local and foreign germplasms for the evaluation and selection of cultivars tolerant to the productivity constraints. The long-term approach, launched in 1995, includes breeding for resistance to the productivity constraints using conventional breeding methods and genetic engineering. Genetic engineering projects in Uganda target the most popular and infertile cultivars that cannot be improved through conventional (cross) breeding. The main objective of genetic engineering in Uganda is to develop genetically modified (GM) cultivars that are resistant to local pests and diseases, have improved agronomic attributes, and are acceptable to consumers (Kikulwe et al. 2007).

The introduction of a GM banana in Uganda is not without controversy. In Uganda, where the technology of genetic engineering is still in its infancy, it is likely to generate a wide portfolio of concerns, as it has in other African countries. According to the Uganda National Council of Science and Technology (UNCST) (2006), the main public concern is the safety of the technology for the environment and human health.

Several countries have designed and implemented policies to address the safety concerns of consumers and producers (Beckmann, Soregaroli, and Wesseler 2006a,b). Such policies include assessment, management, and communication of the biosafety profiles of genetically modified organisms (GMOs) (Falck-Zepeda 2006). As a consequence of its international obligations and the need to guarantee a socially accepted level of safety to its citizens,

Uganda has taken significant steps to ensure the safety of GM biotechnology applications. GM banana varieties will need to undergo biosafety assessments[1] and receive the regulatory approval of the country's National Biosafety Committee before being approved for research, confined field trials, and release into the environment for commercialization.[2]

However, the biosafety regulatory process has several economic consequences, as biosafety regulations are not costless endeavors. Kalaitzandonakes, Alston, and Bradford (2007) calculate the compliance costs for regulatory approval of herbicide-tolerant and insect-resistant (Bt) maize to be on the order of about 7–50 million US dollars (US$). They note that the approval costs for similar types of GM crops will be alike. In addition, biosafety-testing requirements can consume significant amounts of time—from a few months to several years. A delay in the approval of a new variety forestalls access to the potential benefits generated by farmer adoption of the technology, and one can expect such costs to be substantially higher than the regulatory compliance costs (Wesseler, Scatasta, and Nillesen 2007).

Jaffe (2006) has noted that existing drafts of Uganda's biotechnology and biosafety policy stress the importance of the socioeconomic implications of the technology for biosafety regulation, but that author also observes a lack of precision in identifying the socioeconomic aspects and how they should be considered. In fact, Article 26.1 of the Cartagena Protocol on Biosafety to the Convention on Biological Diversity gives countries the choice of whether to include socioeconomic considerations in the biosafety assessment process consistent with other international treaties, although limited to the context of biodiversity (Jaffe 2006). Article 26.1's "may take into account" clause has been applied strictly in some countries, such as India, where the socioeconomic consideration is mandatory for biosafety applications.

Many countries, including Uganda, have not determined whether and how to include socioeconomic considerations, at what stage of the regulatory process to include them, and what the scope and decisionmaking process within biosafety regulations should be. In fact, some biosafety experts (and some countries) have resisted including a socioeconomic assessment as a mandatory part in the biosafety decisionmaking process, as in their view, such issues may

---

1   The original scope of biosafety as described in the Cartagena Protocol on Biosafety was environmental safety. However, over time the original scope has been expanded to include food and feed safety in terms of toxicity or allergenicity. In this chapter, it is therefore understood that the label biosafety includes both environmental and food/feed safety.

2   Five technologies have been approved for confined field trial testing in Uganda: a virus-resistant cassava, a weevil-resistant sweet potato, an insect-resistant and herbicide-tolerant cotton, a fungus-resistant banana, a bacteria-resistant banana, and a nutrition-enhanced banana.

cloud that process and distract regulators from the scientific/technical issues related directly to biosafety. It is worthwhile to note that inclusion of socio-economic considerations for biosafety regulatory approval at the laboratory/greenhouse or confined field trial stages contributes very little to the decision-making process, as the material will not enter the food chain and thus will not be commercialized until it is given regulatory approval further along in the process. Therefore, a major objective of this chapter is to illustrate the relevance of socioeconomic analyses for supporting biotechnology decisionmaking (and in particular, the importance of consumer perceptions) but also for contributing to the development and implementation of biosafety regulations. We present a general approach using the GM banana as an example, assuming the GM banana has passed standard food safety and biosafety assessments and thus can be considered to be safe.

The GM banana is the pioneer staple foodcrop in Uganda developed through modern biotechnology. Though different transgenic traits are being developed (see Shotkoski et al. 2010), in this chapter our focus is on a fungus-resistant GM banana with a trait resistant to the airborne fungal leaf-spot disease known as black Sigatoka (*Mycosphaerella fijiensis*), which can reduce yields by 30–50 percent. The development of an agronomic trait possessing this fungal disease resistance is important, as the new banana can substantially increase yields, which would directly improve the livelihoods of farmers (Kikulwe, Wesseler, and Falck-Zepeda 2008; Kikulwe 2010). The bananas targeted for modification are the East African endemic cultivars, including cooking bananas that are mostly grown and highly preferred by consumers.

Given the importance of the crop, a better understanding of the socio-economic effects of introducing a GM banana is desirable to build public confidence in the technology and its implications for food security. In addition, assessing the potential benefits as well as the economic welfare will shed light on the question of under which conditions will Uganda in particular, and African countries in general, gain from GM crops without making a particular population segment worse off. Specifically, five sets of research questions are addressed in this study:

1. What are the expected social incremental benefits and costs under the conditions of irreversibility, flexibility, and uncertainty of introducing GM bananas in Uganda?

2. What are consumers' knowledge about and attitudes and perceptions toward introducing GM bananas in Uganda? How do they differ between rural and urban households? Do consumers know, and have trust in, the

institutions responsible for regulating, releasing, and selling GM crops in Uganda?

3. How does preference heterogeneity influence choice of banana bunch attributes across individual households? What are the differences between consumer preferences in urban and rural households for banana bunch attributes?

4. How much are consumers willing to pay for the values accruing from GM bananas given a social benefit? How does this willingness compare across different segments of consumers?

5. What are the impacts of introducing GM bananas on food security in Uganda? What implications does introduction have for biosafety regulations in general?

In the following sections, we discuss the benefits that a GM banana could provide to producers and consumers in Uganda and the role of biosafety regulations in governing the introduction of a GM banana. The results of a real option model are presented that show how concerns about environmental risks can be considered in a cost-benefit analysis as a first step toward a socioeconomic assessment of introducing a GM banana in Uganda.

In addition, we show how the results of the economic analysis can be combined with the consumers' willingness to pay (WTP) for a GM banana using a choice experiment model. We explicitly demonstrate how one can use both random-parameter logit and latent-segment models to capture and account for heterogeneity among consumer preferences given a tangible economic benefit of the GM banana.[3] The study complements and extends the dimensions of previous research (Li et al. 2003; Loureiro and Bugbee 2005; Knight et al. 2007) on consumers' WTP for GM food by, first, incorporating the forgone economic benefits of a delay in release and, second, incorporating producers as consumers in the sample. The approach is unique in its application to banana varieties in a developing-country context.

Before presenting the results of the WTP analysis, results of consumer knowledge, attitudes, and perceptions (KAP) toward the GM banana and its regulation are introduced. The consumer perceptions presented in this chapter hold numerous implications for scientists, policymakers (regulators), the public, and other stakeholders.

---

3 A tangible economic benefit refers to a benefit forgone if the GM banana is not introduced immediately.

The aim of this chapter is to mention and discuss a range of these implications. The contribution is structured as follows. The next section discusses in more detail the relevance of a GM banana for Uganda. The following section introduces an overview of biosafety regulations in Uganda. The subsequent section presents the overall approach and explains its application. The main results are then reported, followed by a section on the policy implications of the empirical findings for decisionmaking on biotechnology and biosafety regulations in Uganda for the GM banana in particular, and other GM crops in general. Limitations and suggestions for future research are discussed in the final section.

Relevance of a GM Banana for Uganda

Banana is one of the most important crops in Uganda, with approximately 7 million people, or 26 percent of the population, depending on the plant as a source of food and income. Bananas are estimated to occupy 1.5 million hectares of the total arable land, or 38 percent of the cultivated land, in the country (Rubaihayo 1991; Rubaihayo and Gold 1993). The plant is grown primarily as a subsistence crop in rural areas, although consumption is not limited to rural areas, as approximately 65 percent of urban consumers in Uganda have a meal of the cooking variety of banana at least once a day. Ugandans have the highest per capita consumption of cooking bananas in the world (Clarke 2003).

Most banana varieties grown in Uganda are endemic to the East African highlands—a region recognized as a secondary center of banana diversity (Stover and Simmonds 1987; Swennen and Vuylsteke 1991; Smale and Tushemereirwe 2007). The endemic banana varieties (AAA–EA genomic group) consist of two use-determined types: cooking bananas (matooke) and beer bananas (mbidde). Karamura (1998) recognized 238 names of East African highland banana varieties in Uganda, with 84 clones grouped into five clone sets. The nonendemic clones include dessert bananas (varieties that are consumed raw), some beer bananas (varieties suitable for beer and juice making), and roasting bananas (or plantains).

Banana yields in Uganda are severely reduced by several pests and diseases. Among the pests that cause the most yield damage are weevils (*Cosmopolites sordidus*) and nematodes (*Radopholus similis, Pratylenchus goodeyi,* and *Helicotylenchus multicinctus*). The diseases that contribute to the worst yield losses in Uganda are the soilborne fungal Panama disease, or *Fusarium* wilt (*Fusarium oxysporum*); bacterial wilts, including the banana *Xanthomonas* wilt (*Xanthomonas campestris* pv. *musacearum*); and the airborne fungal leaf-spot disease black Sigatoka (*Mycosphaerella fijiensis* Morelet) (Gold 1998, 2000; Gold et al. 1998; Gold, Pena, and Karamura 2001; Tushemereirwe et al. 2003b).

Consequently, the National Banana Research Program of the National Agricultural Research Organisation (NARO) in Uganda has developed a breeding program that employs a range of traditional crop-breeding methods and a portfolio of biotechnologies to address the crop's most debilitating problems caused by pests and diseases (Kikulwe et al. 2007). The short-term breeding strategy includes the assembly of local and foreign germplasms for evaluation and selection of varieties resistant or tolerant to existing productivity constraints. Resistance to a limited set of pests and diseases (for example, black Sigatoka) was identified in hybrid banana varieties. Though characterized by bigger bunches, the hybrid varieties are not widely grown in Uganda (Nowakunda 2001; Smale and Tushemereirwe 2007). Producers and consumers prefer the East African highland cooking bananas, but these are also highly susceptible to black Sigatoka (Nowakunda et al. 2000; Nowakunda 2001) and bacterial wilts (Tushemereirwe et al. 2003a). Susceptibility to diseases prompted the national researchers to adopt a long-term breeding strategy that includes the generation of new genotypes and other new approaches to introduce resistance.

The highest yielding highland cooking bananas proved to be sterile, which slows down their improvement through conventional breeding (Ssebuliba 2001; Ssebuliba et al. 2006). With major biotic constraints not easily addressed through conventional breeding and management practices, recent efforts have been made to employ genetic engineering for the insertion of resistance traits into selected banana background planting material. Unlike cross-breeding, genetic engineering allows for improving the agronomic traits (for example, disease and pest resistance), as genes are inserted into potential host varieties (cultivars) while not changing other production and product attributes (for example, cooking quality). The genetic modification approach has shown potential for the improvement of the crop (Tripathi 2003).

Edmeades and Smale (2006) argue that the choice of a host variety for a genetic transformation largely determines its acceptability by producers and consumers. In those regions strongly affected by biotic constraints, it is likely that GM banana cultivars will be more beneficial to poorer and subsistence-oriented farmers. In addition, the insertion of multiple traits into East African highland bananas, although associated with additional R&D costs (for example, transformation and regulatory costs), could further increase the benefits generated by the adoption of the technology in Uganda. Multiple traits may also increase adoption rates, as farmers may not immediately notice the beneficial effect of a single trait.

Although GM bananas look promising for large-scale (mass clonal) multiplication and dissemination, empirical evidence of the success of such organisms is still limited. Long-term multiplication of micropropagated (tissue-cultured) plants, for example, may lead to epigenetic[4] (somaclonal) variations. Additionally, genetic uniformity in a trait intensifies the probability of mutations in the targeted pest or disease that overcome resistance and increase epidemic vulnerability. These two aspects raise questions about the clonal fidelity of offspring plants and their genetic stability, both affecting economic benefits of GM banana varieties. In this context biosafety measures to monitor, evaluate, and mitigate effects of such occurrences become critical for the appropriate deployment of the technology in Uganda.

Despite these possible effects on the persistence of economic benefits of a GM banana, it is important to note that throughout the chapter we assume the GM banana has been proven safe for human health and the environment according to standard safety assessments.

Biosafety Regulations in Uganda

Uganda is among the few African countries that have invested in agricultural GM crop R&D and have initiated procedures for confined field trials to evaluate GM technologies (Atanassov et al. 2004).[5] The country has taken significant steps to ensure safety in biotechnology application (Nampala, Mugoya, and Ssengooba 2005). Biosafety regulations and, to a degree, biotechnology developments in Uganda are governed in the context of the Cartagena Protocol on Biosafety (GOU 2002b, 2004). Uganda signed the protocol in May 2000 and ratified it in November 2001 (GOU 2004; Wafula and Clark 2005). UNCST is the institution responsible for implementation of the biosafety protocol and is the protocol's designated competent authority. UNCST

---

4 Epigenetic changes are changes that do not affect the DNA sequence of genes but change the gene in other ways. These changes may be induced spontaneously in response to environmental factors or to the presence of a particular allele, even if it is absent from subsequent generations. Modgil et al. (2005) note that the in vitro process, length of in vitro culture, and in vitro stress stemming from unnatural and nutritional conditions are some of the factors believed to induce epigenetic changes.

5 The GM banana field trials were approved by the National Biosafety Committee and have been established at Kawanda by NARO. Note that Uganda now joins the other five African countries that have conducted confined field trials of GM crops: Burkina Faso, Egypt, Kenya, South Africa, and Zimbabwe. Of those, only Burkina Faso, Egypt, and South Africa have approved crops for commercialization.

established the National Biosafety Committee (NBC), a technical evaluation arm, in 1996. NBC is responsible for reviewing applications and implementing general biosafety guidelines and regulations (GOU 2004; Wafula and Clark 2005).

Currently, the basis for the development and application of biotechnology is the National Science and Technology Policy of 2001 (GOU 2004). The National Science and Technology Policy provides general reference to biotechnology within the broader context of the role of science and technology in national development. The responsibilities of the various institutions and agencies involved in the approval process for biotechnology products have been outlined in the Biosafety Framework of 2000. The National Biosafety Framework, which was developed by UNCST, is based on the United Nations Environment Programme's International Technical Guidelines on Safety in Biotechnology (enacted in 2001). Those guidelines provide terms of reference for NBC and the institutional biosafety committees—detailed information on risk assessment and management procedures for microbes, inspection, and approval (Traynor 2003).

Under the National Biosafety Framework, UNCST has the mandate of approving GMOs for research purposes, confined release into the environment, and commercial planting in Uganda. UNCST receives all applications for research on or the deliberate introduction of GMOs, conducts a screening for completeness, and enters the applications into the national public records as submitted, before forwarding them to NBC for review and risk assessment evaluation. The complete risk assessment is done by the applicant. The NBC is obliged to review the risk assessment dossiers submitted by the applicant after the application has been assessed by the institutional biosafety committees and finally advises UNCST.

The members of NBC are stakeholders, such as representatives from regulatory agencies, the scientific community, universities, the private sector, and civil society (Nampala, Mugoya, and Ssengooba 2005). The NBC is also responsible for writing the draft National Biotechnology and Biosafety Policy, the draft National Biosafety Regulations (GOU 2004), and the Guidelines on Biosafety in Biotechnology for Uganda (GOU 2002a), and for developing draft manuals addressing specific issues surrounding biosafety regulations, such as confidential business information. These documents make up the biosafety regulatory framework for Uganda. At this time, the documents still need to be approved by the government (Jaffe 2006). The implementation of the finally agreed-upon biosafety regulations for a specific GM crop will be managed by UNCST. UNCST is advised by the National Biotechnology

Advisory Committee, which is an interministerial committee including representatives from key institutions in biotechnology development and NBC (GOU 2004).

Applications for the import and export of GMOs are also approved by NBC. The government of Uganda (GOU 2005, 6) stipulates that "any person, prior to intentionally introducing a GMO shall apply to the Competent Authority for authorization. In case of imports, the exporter or the Competent Authority in the country of export may submit an application on behalf of an applicant, and may designate in the application with whom the Competent Authority shall communicate regarding the application." In the case of field testing, the government of Uganda (GOU 2005, 7) continues to specify that "the applicant shall document to the Competent Authority that participating personnel will have appropriate training [and] that the field test will be overseen by an individual possessing appropriate technical expertise." The competent authority has to reply within 90 days. Within a period of not more than 270 days after the scientific risk assessment, the competent authority will make a final decision on whether to approve or deny the applicant the authority to introduce the GMO. Thus, if all the required documents are submitted and complete, the approval will take about one year. In the case of a denial of a request, the applicant is given 30 days to appeal with genuine reasons or additional relevant information, and the competent authority has 30 days to render a final decision. The competent authority will finally make public any proposal about the intentional introduction of GMOs. The public is given not more than 60 days to submit comments, which the competent authority will take into consideration before the final decision is made.

According to Wafula and Clark (2005), NARO submitted applications to UNCST in 2000 to introduce Bt cotton and Bt maize, but they were not approved for confined field trials. One of the reasons UNCST gave was that Uganda was unprepared to handle GM crops, because it lacked a national biotechnology and biosafety policy. In addition, Uganda lacked confinement and containment facilities for GMO field trials. Recently, the government selected biotechnology as one of the priority areas in its plan for the modernization of agriculture (OPM 2005). Consequently, substantial investments have been made in institutional development and capacity building for agricultural biotechnology and biosafety.

Jaffe (2006) analyzed and compared biosafety regulatory systems in Uganda, Kenya, and Tanzania using the African Model Law, which is one of the documents given consideration throughout Africa when a country begins drafting laws and regulations to address biosafety. Jaffe's assessment considered key issues, such as comprehensiveness, transparency, participation, and efficiency of the regulatory systems. In the context of Uganda's existing biosafety regulatory framework, despite all the efforts that have been made, the author notes some shortcomings, particularly in the areas of transparency and clarity regarding the process to be followed, so that all interested stakeholders are able to understand and meet the requirements of the regulatory process. Those shortcomings include, first, no clear indications of how the assessment of the potential food-safety risks that might arise from the GMOs will be handled. Second, even though UNCST is involved in the formulation of the biosafety regulation policy, the statute authorizing its creation does not provide legal authority to regulate GMOs. Neither the biosafety policy nor the government regulations establish a clear safety standard for approving a GMO. Third, the documents contain no elaborations on how and what socioeconomic considerations will be considered, how they will be analyzed, by whom, and how they will be considered in the decisionmaking process.

This chapter seeks to address the third shortcoming by providing information that, indeed, socioeconomic issues of introducing GMOs in Uganda are relevant. It also presents methods to identify socioeconomic issues and provides suggestions for addressing them.

Approach and Implementation

The framework of the research comprises two approaches: real options and a choice experiment. The latter relates the economic benefits to potential consumer concerns. Primary and secondary data sources were used for this study. Primary data were generated from a survey conducted in three administrative regions, Eastern, Central, and Southwestern Uganda, comprising three distinct agroecological zones where cooking bananas (green bananas) are produced and consumed. The study was implemented in July and August 2007 with face-to-face interviews. Six enumerators were hired and trained specifically for this study. In implementing the survey, enumerators briefly described the context of the study and informed the respondents that there were no wrong or right answers but their opinions were of interest. A total of 421 households, drawn with a random sample stratified into rural and urban households using the then-current community listings of 21 randomly selected communities, allowed us to draw general conclusions (Figure 4.1).

The survey questionnaire was designed to collect information on the respondents' observed characteristics. First, each respondent was asked questions about his/her KAP regarding GM crops and food. In part two, social,

FIGURE 4.1   Location of study sites

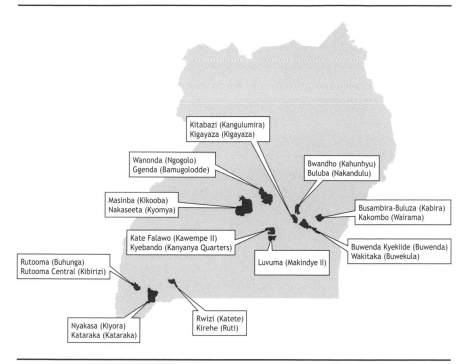

Source: Authors.

demographic, and economic information on the households was collected, including the characteristics of the banana purchase, decisionmaker(s), and other members of the household. The final part consisted of the choice experiment. The detailed description of the study area, the questionnaire, and the information provided to the respondents is reported in Kikulwe (2010). Secondary data are taken from the database of a NARO/International Food Policy Research Institute project conducted between 2003 and 2004 in Uganda. The dataset is complemented by data for banana production for 1980 through 2004 obtained from the Uganda Bureau of Statistics (UBOS 2006) and the Food and Agriculture Organization of the United Nations (FAO 2006).

Different econometric models were applied to the datasets to test hypotheses related to the five research questions. A real option model was used to estimate the maximum incremental socially tolerable irreversible costs (MISTICs) for GM bananas, providing a maximum threshold value for consumer perceived irreversible costs of introducing GM bananas, as explained in more detail in Kikulwe (2010) and Kikulwe, Wesseler, and Falck-Zepeda (2008).

An explanatory factor analysis was applied to investigate the underlying latent structure of the KAP data. Random-parameter logit models and latent class models were then applied to investigate consumers' preference heterogeneity for banana attributes in the choice data. Finally, we compared the MISTICs with the willingness to pay for GM bananas to derive policy implications.

Overview of Findings

To achieve the overall aim of this study, five research questions were addressed. This section presents the highlights for each research question.

1. What are the expected social incremental benefits and costs under the conditions of irreversibility, flexibility, and uncertainty of introducing GM bananas in Uganda?

A real option approach[6] (see Appendix 4A) was followed to analyze the social incremental benefits and costs (that is, the MISTICs) and social incremental reversible benefits (SIRBs). Irreversibilities and uncertainties have been considered in the literature on introducing GM crops (for example, Wesseler, Scatasta, and Nillesen 2007). Scatasta, Wesseler, and Demont (2006) introduced the term MISTICs to identify the threshold value for consumers' WTP for not having a GM crop introduced. The MISTICs associated with the adoption of a GM banana in Uganda were calculated using equation 4A.1 in Appendix 4A and presented for different risk-free and risk-adjusted rates of return, as shown in Table 4.1. The results showed the MISTICs to be between approximately US$176 million and US$359 million per year, or between US$282 and US$451 per hectare per year. In the scenario with a risk-adjusted rate of return of 12 percent and a risk-free rate of interest of 4 percent—which we considered to be a reasonable scenario based on the results of Mitthöfer (2005)—the annual MISTICs per household are about US$38. This result can be interpreted as follows: the immediate release of the GM banana should be postponed or abandoned only if the average

---

6 A real option approach considers the irreversible effects to see how the stream of incremental benefits will be affected over a long planning horizon (30 years or more)—with continuous state and continuous time.

TABLE 4.1 Hurdle rates and average annual MISTICs per hectare of genetically modified bananas, per household, and per banana-growing farm household at different risk-free rates of return and risk-adjusted rates of return

Risk-free rate of return (r)		Risk-adjusted rates of return (μ)					
		0.04	0.06	0.08	0.10	0.12	0.14
0.00	Hurdle rate	1.0169	1.0104	1.0075	1.0059	1.0048	1.0041
	MISTIC (million US$)	359	301	258	225	199	178
	MISTIC (US$/ha)	451	394	353	324	302	285
	MISTIC (US$/household)	69	58	50	43	38	34
	MISTIC (US$/farmer)	239	201	172	150	133	119
0.04	Hurdle rate	1.3298	1.0405	1.0166	1.0103	1.0075	1.0058
	MISTIC (million US$)	274	293	256	224	198	178
	MISTIC (US$/ha)	345	383	350	322	301	285
	MISTIC (US$/household)	53	56	49	43	38	34
	MISTIC (US$/farmer)	183	195	170	149	132	119
0.10	Hurdle rate				1.1386	1.0355	1.0161
	MISTIC (million US$)				199	193	176
	MISTIC (US$/ha)				286	293	282
	MISTIC (US$/household)				38	37	34
	MISTIC (US$/farmer)				132	129	118

Source: Calculation by authors.

Note: The exchange rate used is US$1 – UGX1,750, for the year 2007. MISTIC = maximum incremental socially tolerable irreversible cost; UGX = Ugandan shilling; US$ = US dollar.

household is willing to give up more than US$38 per year for not having such a banana introduced.

In the case where approval of the GM banana is delayed due to missing regulatory procedures and protocols, Uganda will forgo potential benefits (SIRBs) in the approximate range of US$179 million to US$365 million per year. This forgone benefit can be an indicator of how much Uganda can pay to compensate for potential damages. Additionally, the SIRBs provide a clue about the maximum costs farmers would endure to comply with biosafety of about US$303 per hectare. Adopters of the GM banana would not be willing to pay more than US$200 per hectare per year in transaction costs—that is, costs to comply with biosafety regulations, R&D costs, and technology transfer costs. If the average WTP per hectare of a banana-growing household is below the MISTICs, but biosafety regulators are inclined to implement biosafety regulations to address concerns of consumers with a high WTP for not having the GM banana, those additional costs should not exceed US$200 on average per year per hectare planted to GM bananas. Assuming a maximum of 541,530 hectares that may be planted with GM bananas in Uganda, this implies that the maximum total costs to bring the GM banana to Ugandan producers cannot exceed US$108 million. Otherwise, the GM banana is not a viable alternative.

Based on the MISTICs results, it is evident that Uganda loses from not introducing a fungus-resistant GM banana. But only if the average household is willing to give up more than US$38 annually for not having GM bananas introduced should an immediate release be postponed. This analysis demonstrates a relationship between agricultural policy, R&D, technology delivery, and impact, which shows an inverse relationship between stringency (precautionary approaches) and technology delivery. That is, the more stringent the approval process is, the greater will be the potential benefits that are forgone annually, which negatively affects both the scientists and the technology end users.

2. What are the consumers' knowledge about and attitudes and perceptions toward introducing GM bananas in Uganda? How do they differ between rural and urban households? Do consumers know, and have trust in, the institutions responsible for regulating, releasing, and selling GM crops in Uganda?

As little is known about consumer KAP toward the GM banana in Uganda, an explanatory factor analysis was applied to investigate the underlying latent structure of the KAP data. The analysis of KAPs reveals the presence of three categories, including benefit, food-environment risk, and health risk KAPs. The KAP toward GM crops among rural and urban consumers varies, owing to a number of socioeconomic characteristics, suggesting a rural–urban bias. Given quality benefits, consumers are more willing to accept GM bananas, but at the same time they are concerned about the unknown negative effects of the technology. Results show that rural consumers value the quality benefits, whereas urban consumers are more concerned about the safety of the technology (Table 4.2). Education and income have negative effects on GM banana acceptability. Results further indicate that there is a relatively high level of awareness and trust in local leaders and extension workers. Respondents were less aware of UNCST and the Consumer Education Trust, the two main agencies responsible for informing consumers about GM food.

TABLE 4.2   Comparison of KAP scores with consumer characteristics

| | Mean KAP scores | | | | | | | | |
| | Benefit | | | Food-environment risk | | | Health risk | | |
Characteristic	Rural	Urban	Test	Rural	Urban	Test	Rural	Urban	Test
Region									
Central	0.18^a	-0.15^a	*	-0.08^a	0.10^a		0.04^a	0.28^a	*
Eastern	0.34^a	-0.27^a	*	-0.12^a	0.10^a		-0.15^a	0.23^a	*
Southwestern	-0.25^b	0.02^a		0.03^a	0.02^a		-0.22^a	0.02^a	
P-value	0.00	0.45		0.53	0.90		0.32	0.27	
Gender									
Men	0.10^a	-0.39^b	*	-0.08^a	-0.07^b		-0.03^a	0.13^a	
Women	0.11^a	0.06^a		-0.02^a	0.19^a	*	-0.28^b	0.28^a	*
P-value	0.87	0.01		0.60	0.09		0.02	0.31	
Education status									
None	-0.02^a	0.31^a		-0.05^a	-0.31^a		-0.14^a	0.04^a	
Primary	0.15^a	-0.08^{ab}	*	-0.05^a	-0.02^a		-0.21^a	0.10^a	*
Secondary	0.23^a	-0.14^{ab}	*	-0.11^a	0.23^a	*	0.03^a	0.21^a	
College or above	-0.69^b	-0.45^b		0.08^a	0.13^a		0.31^a	0.41^a	
P-value	0.00	0.15		0.89	0.17		0.05	0.40	
Income level									
Low	0.20^a	-0.33^a	*	-0.09^a	-0.22^a		-0.05^a	0.17^a	
Medium	-0.02^b	0.11^a		0.02^a	0.17^a		-0.12^a	0.09^a	
High	0.13^{ab}	-0.24^a	*	-0.15^a	0.14^a		-0.34^a	0.28^a	*
P-value	0.09	0.19		0.48	0.12		0.16	0.48	
At least one family member employed off-farm									
Yes	-0.05^b	-0.06^a		-0.15^a	0.10^a	*	-0.16^a	0.16^a	*
No	0.20^a	-0.48^b	*	-0.001^a	0.01^a		-0.10^a	0.32^a	*
P-value	0.01	0.06		0.16	0.59		0.57	0.36	
Banana status									
Grow only	0.07^{ab}	-0.01^a		-0.12^a	0.003^a		-0.16^a	0.17^a	
Buy only	0.45^{ab}	-0.21^a	*	0.07^a	-0.01^a		-0.23^a	0.28^a	*
Grow and sell	-0.05^b	-0.12^a		0.02^a	0.46^a		0.02^a	0.32^a	
Grow and buy	0.27^a	-0.02^a	*	-0.10^a	0.05^a		-0.33^a	0.13^a	*
Grow, sell, and buy	-0.02^{ab}	-0.19^a		-0.10^a	0.38^a		0.11^a	-0.26^a	
P-value	0.03	0.91		0.89	0.39		0.03	0.30	

Source: Authors.

Notes: * denotes significance at the 10 percent level or better. In columns, means followed by the same superscript letter are not significant at the 10 percent level or better (Sidak multiple-comparison test in STATA). (This note refers to the similarly lettered superscripts on means in columns. For instance, under gender, the benefit KAP scores for men and women respondents in the rural areas were not significantly different: they both carry superscript a.) KAP = knowledge, attitudes, and perceptions.

In conclusion, we argue that delaying the approval of a fungus-resistant GM banana in Uganda is more in line with the preferences of urban (in particular, the better educated and wealthier) consumers than with rural ones. But how can the negative perceptions among the urban and wealthier ones become positive or neutral at best? There is a need to ensure transparency and participation but to strike a balance with the feasibility of a system. However, if the system is not participatory and does not respect dissenting opinions, then legitimacy is taken from it. If this is the case, then people tend not to respect the regulatory system. The main lesson is for NARO and the government of Uganda to develop in advance communication strategies to ensure proper discussion and address potential concerns.

3. How does preference heterogeneity influence choice of banana bunch attributes across individual households? What are the differences between consumer preferences in urban and rural households for banana bunch attributes?

The heterogeneity in consumers' preferences for different banana attributes in Uganda was investigated using choice experiment data. The analysis of the choice data took into consideration preference heterogeneity resulting from locational and household-level characteristics. This helped to test whether consumers in rural and urban locations value banana attributes differently. A random-parameter logit model[7] was applied to investigate the heterogeneity preference for the banana bunch attributes. Interactions of respondent-specific household characteristics with choice-specific attributes in the utility function were included in the model to account for the source of unobserved heterogeneity. This provided insights about differences in consumer valuation of the GM technology in addition to explaining the aggregate economic value associated with such technology, similar to the policy-change effect as analyzed by Boxall and Adamowicz (2002). Recent applications of random-parameter logit models (for example, Breffle and Morey 2000; Carlsson, Frykblom, and Liljenstolpe 2003; Kontoleon 2003; Morey and Rossmann 2003) have revealed that this model is superior to the conditional logit

---

7 The random-parameter logit model estimates both the mean coefficient and standard deviation of the random parameters. It is imperative to note the following. First, if the standard deviation estimate is not significantly different from zero, then one can conclude that the preference parameter is constant across the population. Second, if the mean coefficient is zero (or significantly smaller than the standard deviation) with a significant estimated standard deviation, then there is preference diversity (that is, both positive and negative). Third, if both the mean coefficient and estimated standard deviation are insignificant, then the attribute has no impact on choices.

model in terms of overall fit and welfare estimates. In this chapter, therefore, a random-parameter logit model was applied, as shown in Appendix 4B. This was followed by a random-parameter logit model including interactions of respondent-specific characteristics with banana bunch attributes to provide more information about the sources of variation in preferences across respondents (see equation 4B.3). The selected respondent-specific characteristics included (1) household size (HHSIZE); (2) whether or not the respondent had postsecondary education (EDUC); (3) log of household monthly income (INCOME); (4) age of the respondent (AGE); (5) whether or not the household grows bananas (GROW); (6) whether or not the household was found in the Eastern region (EAST); and (7) whether or not the household was found in the Southwestern region (SWEST). Findings reveal that there is substantial conditional and unconditional heterogeneity, as accounted for by the random-parameter logit model with interactions, carried out for each location (urban and rural) separately (Table 4.3).

The impacts of social and economic characteristics of the consumers on their valuation of the banana bunch attributes were significant, indicating the importance of considering such characteristics in explaining the sources of conditional heterogeneity. Even though bunch size is valued highly by both rural and urban households, urban and rural preferences differ concerning the introduction of a GM banana. The low-income rural households with larger household sizes value the GM technology that generates benefits to producers more highly than do the urban ones. Conversely, respondents with higher education were found to be more critical of the GM technology, which would negatively influence their willingness to accept the GM banana. Statistical tests confirm that there are significant differences in preferences for banana bunch attributes between urban and rural households in Uganda.

The application of the econometric models in this section supports two conclusions. First, a connection needs to be established between banana attributes and crop improvement efforts. In that sense, there is a need to link plant breeders, consumers, producers, and decisionmakers. For instance, it is evident in the findings that bunch size matters a lot for both rural and urban respondents. Therefore, breeding efforts should concentrate on improving bunch size but without forgetting other quality attributes. Second, increasing the participation of consumers, producers, and producers who happen to be consumers in the decisionmaking process and in marketing chains can help reduce negative perceptions. This is important, not only because of the benefits, but also because of negative responses (such as anti-GM banana campaigns) that failure to include these stakeholders may trigger.

TABLE 4.3   Random-parameter logit model with interactions

	Rural consumers		Urban consumers	
	Coefficient	Coefficient standard deviation	Coefficient	Coefficient standard deviation
Random parameters in utility function				
GM biotechnology	1.02*** (0.20)	1.27*** (0.13)	1.56*** (0.29)	1.75*** (0.20)
Large benefit	0.28*** (0.10)	0.41** (0.21)	−0.75* (0.41)	0.06 (0.26)
Nonrandom parameters in utility function				
ASC	1.34*** (0.13)		1.02*** (0.17)	
Medium bunch size	0.24*** (0.06)		0.65*** (0.05)	
Large bunch size	0.37*** (0.08)		0.89*** (0.07)	
Medium benefit	0.13*** (0.03)		0.13*** (0.05)	
Price (% change)	−0.02*** (0.00)		−0.02*** (0.00)	
Medium bunch size × EDUC	−0.18* (0.10)		−0.27*** (0.10)	
Medium bunch size × HHSIZE	0.03*** (0.01)			
Large bunch size × HHSIZE	0.04*** (0.01)			
Large bunch size × GROW			0.64*** (0.15)	
Large bunch size × EDUC			−0.54*** (0.12)	
Large benefit × EDUC	−0.30** (0.12)			
Large benefit × INCOME			0.15** (0.07)	
GM biotechnology × EDUC	−0.40** (0.16)		−1.10*** (0.19)	
GM biotechnology × HHSIZE	0.06*** (0.01)		−0.10*** (0.02)	
GM biotechnology × AGE	−0.01*** (0.00)			
GM biotechnology × EAST	0.32*** (0.11)		−0.73*** (0.16)	
GM biotechnology × SWEST	−0.52*** (0.10)			
Log likelihood at start	−3,334.05		−1,781.47	
Simulated log likelihood	−3,285.82	$N = 4,496$	−1,729.13	$N = 2,240$
Likelihood ratio test	$96.5(\chi^2_{0.99}(18)) = 34.8$		$104.7(\chi^2_{0.99}(16)) = 32.0$	
	McFadden's $\rho^2 = 0.333$		McFadden's $\rho^2 = 0.284$	

Source: Authors.

Notes: Standard errors are in parentheses. * denotes significance at the 10 percent level, ** significance at the 5 percent level, and *** significance at the 1 percent level. Replications for simulated probability were 500. AGE = age of the respondent; ASC = alternating specific constant; EAST = whether or not the household is in the Eastern Region; EDUC = whether or not the respondent has postsecondary education; GM = genetically modified; GROW = whether or not the household grows bananas; HHSIZE = household size; INCOME = log of monthly household income; SWEST = whether or not the household is in the Southwestern region.

4. How much are consumers willing to pay for the values accruing from GM bananas given a social benefit? How does this willingness compare across different segments of consumers?

Unlike the previous section, where models with interactions and split samples were used to explain heterogeneity of preferences at the individual level, to answer this question we employed a latent class model, which is a more recent model to investigate preference heterogeneity. The latent class model has successfully identified the sources of heterogeneity at the segment level, unlike the covariance heterogeneity models and random-parameter logit models, which capture heterogeneity at the individual level. Investigation of heterogeneity at the segment level would be most policy relevant when assessing the welfare impact of the introduction of a technology, such as a GM food product, on different segments of the population (see, for example, Hu et al. 2004; Kontoleon and Yabe 2006; Birol, Villaba, and Smale 2009). This approach depicts a population as consisting of a finite and identifiable number of segments, or groups of individuals. Preferences are relatively homogeneous within segments but differ substantially from one segment to another. The number of segments is determined endogenously by the data. The fitting of an individual into a specific segment is probabilistic and depends on the social, demographic, and economic characteristics of the respondents, as well as on their KAPs. Furthermore, respondent characteristics affect choices indirectly through their impact on segment membership.

An increasing number of studies have used this approach to estimate farmers' and consumers' preferences for various agricultural technologies and food items. For example, Scarpa et al. (2003); Ouma, Abdulai, and Drucker (2007); and Ruto, Garrod, and Scarpa (2008) employed this model for the valuation of livestock attributes. Hu et al. (2004); Owen, Louviere, and Clark (2005); and Kontoleon and Yabe (2006) used it to investigate consumer preferences for GM food. And Birol, Villaba, and Smale (2009) used it to examine farmer preferences for agrobiodiversity conservation and GM maize adoption.

This analysis involved, first, testing further whether preferences of urban households differ from those expressed by rural households. Second, this study included welfare benefits for producers as one of the attributes. Producer benefits often have not been considered in studies on consumer preferences regarding GM food, and we expect these to have a positive effect on consumers' preferences, similar to the results reported by Loureiro and Bugbee (2005) and Gaskell et al. (2006). The theoretical approach for investigating preference heterogeneity using the latent class model is briefly

summarized in Appendix 4C; a detailed explanation and empirical results
are reported in Kikulwe et al. (2011).

The findings show that there is significant heterogeneity in consumer preferences in our sample. The analysis identified two distinct segments of banana
consumers, the *potential GM banana consumers* (representing 58 percent of the
sample and residing more often in rural areas), and the *potential GM banana
opponents* (representing 42 percent of the sample, with the majority found in
urban areas). GM bananas are valued the most by poorer households who are
located in the rural areas of the Eastern region, where banana pests and diseases are prevalent. These consumers are also younger and have positive opinions regarding the benefits of GM food and crops. They have larger families
and are less often employed off-farm, and they have relatively lower monthly
incomes. They would be willing to pay larger premiums for GM bananas
and to ensure producers of bananas derive higher benefits (Table 4.4). The
empirical findings support Edmeades and Smale (2006), who argue that cli-
ents of GM banana-planting materials are likely to be the poorer, subsistence-
oriented households in regions greatly affected by biotic pressures. These
results are also consistent with results for the second and third research questions. The utility of the *potential GM banana opponents'* segment, mainly representing urban consumers, decreases with the introduction of a GM banana,
which generates benefits to producers. These consumers would therefore be
willing to accept a discount for both GM bananas and benefits to producers. Most of these consumers are older and better off; they reside mainly in

TABLE 4.4   Segment-specific valuation of banana bunch attributes (percentage change in price per banana bunch)

Banana attribute	Segment 1: Potential GM banana consumers (*N* = 285)	Segment 2: Potential GM banana opponents (*N* = 176)	Weighted average (*N* = 421)
Medium bunch size**	31.1 (27.5, 35.1)	37.7 (25.1, 57.6)	33.8 (26.5, 44.5)
Large bunch size***	43.1 (38.7, 48.2)	56.1 (39.7, 81.9)	48.6 (39.1, 62.3)
Medium benefit***	11.2 (8.5, 14.9)	−20.9 (−24.1, −15.9)	−2.3 (−5.2, 1.5)
Large benefit***	18.1 (14.9, 21.7)	−75.3 (−82.1, −70.9)	−21.1 (−21.9, −21.1)
GM biotechnology***	42.5 (36.6, 49.3)	−62.4 (−81.7, −56.8)	−1.5 (−2.6, −1.3)

Source: Kikulwe et al. (2011).

Notes: Numbers in parentheses are the 95 percent confidence intervals. Consumers' valuation of banana attributes were calculated with the Delta method of the Wald procedure contained in the software program LIMDEP 8.0 NLOGIT 3.0 (Econometric Software, New York). ** denotes significance at the 5 percent level and *** significance at the 1 percent level. GM = genetically modified.

urban areas of the Southwestern and Central regions, and mostly associate the GM banana with risks (that is, food, environmental, and health risks). The total WTP among those who gain (the *potential GM banana consumers*) from the introduction of the GM technology is greater than the total WTP among those who lose (the *potential GM banana opponents*) due to this technology. The findings suggest that the gainers (the majority of whom are rural consumers) can potentially compensate the *potential GM banana opponents* (mostly urban consumers) if a GM banana is introduced in Uganda, which is in accordance with the Hicksian compensation criterion (Just, Hueth, and Schmitz 2004).

The latent class econometric analysis supports several conclusions related to the introduction of GM bananas in Uganda. First, findings confirm that GM bananas could be a potentially pro-poor biotechnology, and their introduction would mostly benefit rural households who grow and buy bananas. Second, we find support for Paarlberg's (2008) argument that negative attitudes of urban elites in African countries can be explained by their views on GM food being closer to the European viewpoint versus that of the rural people in their own country. Our empirical findings suggest that better educated people are on average more strongly opposed to GM bananas, which holds not only in urban areas but also in rural ones. Third, rural consumers are willing to pay a higher premium for producer benefits compared to their urban counterparts, suggesting a significant difference between urban and rural consumers' preferences regarding producer benefits. But findings indicate that stressing the potential benefits the technology may provide to farmers is more likely to increase the opposition toward the GM banana among the urban consumers. Based on the preferences of the various groups of stakeholders, the introduction of GM bananas would be desirable for Ugandan society as a whole, and would merit policy support, albeit with consideration of compensation mechanisms aimed at smoothing the gains and losses of benefits. Finally, the main lesson learned is that if preference heterogeneity of consumers is not considered, then the results are likely to be biased. Therefore, for studies that seek to explore consumer preferences, heterogeneity is the primary hypothesis. This has implications for study design, scope, and selection of best practices for evaluation purposes.

5. What are the impacts of introducing GM bananas on food security in Uganda? What implications does introduction have for biosafety regulations in general?

The empirical findings estimated to answer questions 1, 2, and 4 are inte-grated in an economic welfare analysis to provide an overall assessment of the effects of introducing GM bananas on aggregate welfare. The MISTICs asso-ciated with the immediate introduction of a GM banana were compared with the estimated WTP values for the GM banana for different scenarios. We applied the concept of compensating surplus to consumers' preferences for a GM banana, and conducted simulations based on different combinations of impacts associated with GM-banana-introduction strategies to estimate the consumers' welfare measures, as presented in Appendix 4D (and explained in detail in Kikulwe 2010). Welfare measures were estimated for the best-fit latent class model (see question 4).

The findings showed that there are respondents who gain and those who lose from the introduction of a GM banana, which is consistent with results in the previous section. The total welfare for those who gain (*potential GM banana consumers*) is greater than the total welfare for those who lose (*poten-tial GM banana opponents*) (Table 4.5).

The *potential GM banana consumers*, who are mostly located in rural areas, acknowledged much higher willingness to pay for all the proposed GM banana

TABLE 4.5   Compensating surplus and 95 percent confidence intervals for four bunch options[a]

Attribute and segment	Base case: Small bunch	Scenario 1: All medium improvement	Scenario 2: All large improvement	Scenario 3: Large bunch with medium producer benefits	Scenario 4: Medium bunch with large producer benefits
Attribute level					
Bunch size	Small	Medium	Large	Large	Medium
Benefits	None	Medium	Large	Medium	Large
Biotechnology	Traditional	GM	GM	GM	GM
Welfare (UGX/bunch)					
Segment 1 (gainers)[a]	3,000	5,542 (5,179, 5,959)	6,112 (5,707, 6,577)	5,905 (5,515, 6,352)	5,750 (5,370, 6,185)
Segment 2 (losers)[b]	3,000	1,631 (1,325, 2,113)	552 (357, 857)	2,183 (1,762, 2,844)	0.8 (−79.6, 127.0)
Weighted average	3,000	3,900 (3,560, 4,344)	3,777 (3,460, 4,175)	4,341.4 (3,939, 4,878)	3,335 (3,081, 3,335)

Source: Authors.

Notes: Numbers in parentheses are confidence intervals. Exchange rate by July 2007 was US$1 = UGX1,750. GM = geneti-cally modified; UGX = Ugandan shilling; US$ = US dollar.
[a]Also refers to the potential GM banana consumers.
[b]Denotes the potential GM banana opponents as identified under question 5 in the text.

alternatives, particularly a GM banana which is characterized by large bunches and large benefits to producers. With this finding, it is evident that benefits to producers played a significant role in the valuation of the banana bunch attributes. These results imply that if a GM technology can improve crop productivity (and hence increase incomes of the rural subsistence households), that technology would be easily accepted among the rural population segment. Nonetheless, when the *potential GM banana opponents'* households are considered further, their total WTP for the proposed banana improvement scenarios were greater than their estimated average MISTICs per bunch (see Figure 4.2). Based on this finding, we could argue that, on the one hand, the *potential GM banana opponents* are likely to pay more than the threshold value of not having a GM banana introduced in Uganda. On the other hand, the calculated

FIGURE 4.2   Value of welfare and maximum incremental socially tolerable irreversible costs (MISTICs) per bunch at different risk-adjusted discount rates

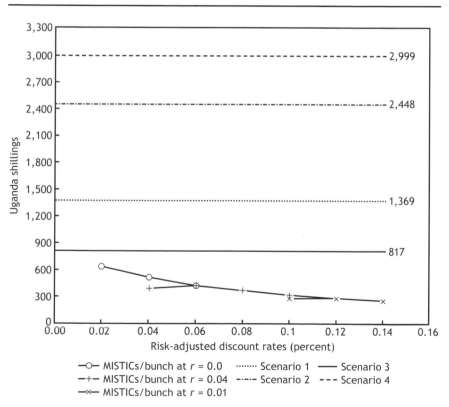

Source: Authors.

MISTICs per bunch for the *potential GM banana opponents'* households were generally low, ranging approximately between US$0.15 and US$0.36. Thus, if the government is to address the concerns of the *potential GM banana opponents*, it will not exceed the potential forgone benefits (US$200) estimated when addressing question 1. There will be still enough to compensate for the negative effects if a fungus-resistant GM banana is introduced. The aggregate welfare showed improvement in welfare over status quo for all scenarios, which is highest when a GM banana with large bunches and medium benefits is proposed. Thus, if a fungus-resistant GM banana with such attributes is now introduced, its introduction may result in strong opposition from the *potential GM banana opponents'* segment of the population, which is composed of mainly urban consumers.

Based on the empirical findings discussed in this section, the following conclusions can be derived. First, the GM banana technology is likely to improve overall welfare in Uganda, but we need to think carefully about those who may lose from the introduction of this technology. Thinking about this beforehand can reduce the loss. But the big question is how do we maximize the benefits and reduce cost and risk? Second, a comprehensive cost–benefit analysis, using different approaches, would be of great importance for assessing the potential benefits and costs of introducing new technologies—such as GM bananas. The net social costs or benefits of most GM crops are likely to be crop-specific, especially in terms of food and environmental safety issues. Introduction strategies would need to consider the distribution of potential costs, benefits, and risk for these new GM crops before a decision to introduce them is made.

Policy Implications

The findings in this chapter demonstrate several implications for different stakeholders in the banana industry in Uganda, and in Africa in general. First, the calculation of the MISTICs considers explicitly possible long-term effects of GM bananas. The results indicate that with each year of delay in the introduction of a GM banana, Uganda loses between about US$179 million and US$365 million. The MISTICs are about US$176 million or more. Only if the real average annual irreversible costs of planting a GM banana would be as high, or higher than, the irreversible benefits, should the release be delayed. We have found no evidence yet that this will be the case. Given the potential and significant economic benefits from the introduction of a GM banana, one might conclude that NARO has to work harder to push the GM banana through the biosafety protocols as promptly and efficiently as possible.

Findings have revealed that government policies delaying the introduction of GM bananas are more in line with the views of wealthier and better educated citizens, the elites, than with the views of the majority of the population. Although this is a disturbing observation—as mainly rural households economically gain from the introduction of a GM banana—a careful approach toward introducing a GM banana is needed to avoid strong urban consumer resistance. In that case, knowing who will be affected by the new innovations is fundamental in foretelling aggregate benefits.

The findings have further shown that the introduction of GM bananas could be beneficial for Ugandan society as a whole and would merit policy support, albeit with consideration of compensation mechanisms aimed at transferring some of the benefits from gainers to losers. Some methods of compensation might be providing more and reliable information about the safety of the technology, which could be channeled through (in addition to the current institutions) local authorities and extension workers. The findings show that there is a high level of awareness and trust in local leaders and extension workers and scant knowledge of UNCST and Consumer Education Trust, the two main agencies responsible for informing consumers about GM food. This finding suggests an opportunity for informing consumers about GM food through local leaders and extension workers. We would recommend instead of UNCST and the Consumer Education Trust informing consumers directly that they use part of their resources for training local leaders and enlisting their help in spreading information. This strategy would help offset the negative KAP toward GM technology, especially among urban consumers.

The approach used here highlights how one can evaluate the socio-economic aspects of GM crops in general, linking both the consumers and adopters of the technology. We have also indicated how one can consider long-term irreversible effects and assess consumer attitudes about GM crops. Empirical research along the lines of the methodology followed in this study can be adapted to new GM crops requiring biosafety assessments prior to commercialization. Such research can help overcome one of the problems of establishing a biosafety system in Uganda and in other developing countries. In particular, NARO may institutionalize the approach suggested in this study and build a system that allows for conducting similar analyses of other GM crops—such as Bt cotton currently undergoing environmental and food safety assessments.

Finally, there is a need to broaden the scope of biosafety processes (now primarily focused on risk) to include food security considerations and agricultural development. This calls for more funding for R&D. Findings have revealed that if a technology has tangible benefits that could improve the incomes of subsistence farmers, that technology could find its way easily to the end users. However, the research agencies that can develop such technologies are financially constrained. For instance, NARO, the main agricultural research agency, which accounts for more than three-quarters of the agricultural research budget in Uganda (ASTI 2002), has received less budget share for fiscal year (FY) 2011/12. That is, the budget for agricultural R&D funded by the government of Uganda has decreased from 12.6 percent of the total agriculture budget in FY2008/2009 to 11.8 percent in FY2011/2012. Similarly, the donor funding for agricultural R&D through NARO has also decreased from 30.0 percent of the total agriculture budget in FY2008/2009 to 17.3 percent in FY2011/12 (GOU 2009, 2012). Yet modern biotechnology was embraced as one of the priority areas targeted by the government to increase incomes and improve the quality of life of poor subsistence farmers through increased productivity and increased share of marketed production (OPM 2005). Our study shows additional financial resources are needed for informing potential opponents about the benefits of the technology, as otherwise resources spent might be wasted.

Limitations and Recommendations for Future Research

The choice experiment approach used to collect data for the model simulations involved mainly the use of surveys of relevant decisionmakers. It involved collection of data from both producers as consumers (who are the likely potential adopters) and those consumers of bananas who do not produce them. The choice modeling technique follows a Lancaster utility approach for analyzing relative importance of product attributes within a relevant product-choice set. However, stated preference approaches are subject to various criticisms. The most important limitation of the choice experiment noted by List and Gallet (2001), akin to other stated preference methods, is that little may be generated from a hypothetical market about the real market behaviors as a result of disparities between hypothetical and actual statements. However, this issue has been addressed by numerous economists in the literature. For example, List, Sinha, and Taylor (2006) compared choice experiments with hypothetical and real situations. In their experiment, the authors informed respondents about the hypothetical bias problem through "cheap talk" and reminded the respondents to take care when making their choices. The authors found no statistically significant differences between hypothetical and real willingness to pay or when estimating the marginal values of attributes. As a result, in our study respondents were informed about the ongoing biotechnological innovations in Uganda using brochures prior to the interviews. They were also reminded that there were no right or wrong answers, and that they should consider their choices carefully. In addition, Lancaster (1966) recommended that to determine the product attribute, it is very important to contact the potential consumers directly. In our study informal interviews with consumers, such as focus group discussions, were used to develop and design the questionnaire, which was later pretested on both rural and urban consumers prior to primary data collection. With a view to the caveats discussed, the findings support Edmeades and Smale (2006)—who used a revealed preference technique to predict the demand of GM banana-planting materials. However, an empirical investigation comparing hypothetical and real market situations may be warranted.

In the empirical analysis of SIRBs, the data for nonprivate net benefits were not available in the public domain. Hence, the SIRBs were estimated based on private net benefits. Furthermore, when estimating the MISTICs, we did not include the transaction costs that might be involved between the technology developers and the end users, including R&D costs, compliance with biosafety regulatory costs, and technology fees.[8] Such costs can be substantial and are one of the major obstacles to technology dissemination in developing countries such as Uganda (Brenner 2004). The problem is not limited to GM technology but includes embodied technologies in general. Adding such costs will reduce the SIRBs. Again, they should on average not be more than the SIRBs per hectare, and should be even less if biosafety regulatory costs at the farm level are added. Another limitation of the study is that the MISTICs calculated were generally for Uganda as a country; however, they are likely to vary by region and even by cultivar. Edmeades (2003) notes the diversity of banana cultivars is high at the country, village, and household levels. On average, 23 different banana cultivars are grown at the village level across Uganda, with approximately 5 different cultivars of cooking bananas grown per household. Households located at high elevations, such as the Southwestern region, were found to grow more cultivars compared to those at low elevations (for example, the Central region and most parts of the Eastern region). Thus, MISTICs may be larger for regions (or households) where banana production

---

8 As technology fees charged by innovators are used to recover R&D and biosafety costs, it is imperative to include such costs as net costs to society to avoid double-counting.

is high compared to those with low banana production. These issues necessitate future empirical research.

Finally, the findings reported in this chapter have shed light on the differences between the urban and rural consumers' preferences regarding banana bunch attributes. However, future research is required to understand in more detail why urban consumers as well as rural and urban elites derive disutility from GM bananas and the associated benefits for producers. In addition, more empirical research is needed to find more mechanisms through which those who gain may compensate those who lose in case GM bananas are introduced.

APPENDIX 4A: MISTICs

We begin with the assumption that incremental reversible net benefits follow a continuous-time, continuous-state process with trend, where GM crops may be released at a point in time. In this approach, the social incremental reversible benefits W^* (or SIRBs; the symbol $*$ indicates optimal timing) need to be greater than the difference between the social incremental irreversible costs (I) and the social incremental irreversible benefits (R), weighted by the size of the uncertainty and flexibility (or hurdle rate) associated with the introduction of the new technology. The hurdle rate is commonly expressed in the form $\beta/(\beta - 1)$, where $\beta > 1$ captures the uncertainty and flexibility effect and is a result of identifying the profit-maximizing decision rule under irreversibility, uncertainty, and flexibility, if benefits do follow a geometric Brownian motion. The geometric Brownian motion is a Wiener process with a geometric trend, for which changes expressed as natural logarithms are normally distributed. The Wiener process is a continuous-time, continuous-state stochastic Markov process with three properties: (1) probability distributions of future values depend on the current value only, (2) the Wiener process grows at independent increments, and (3) changes are normally distributed. The assumption that the adoption of this technology follows a geometric Brownian motion accounts for the uncertainty of the technology (Cox and Miller 1965). The interpretation of the decision rule for the case of a GM banana is that, as long as $W - \dfrac{\beta}{\beta - 1}(I - R) \leq 0$, Uganda should delay adoption of a GM banana until more information about the new technology is available.

In the context of GM crops, where people are more concerned about the not-so-well-known irreversible costs of the technology, it is feasible to estimate threshold values that indicate the maximum incremental social irreversible costs that an individual or society in general is willing to tolerate as compensation for the benefits of the technology. Scatasta, Wesseler, and Demont (2006) have called this threshold value the MISTIC (I^*). In the specific case of Uganda, the estimated MISTICs can be interpreted as the maximum willingness to pay (WTP) for not having the GM banana approved for planting in the country. Actual incremental irreversible social costs (I) are to be no greater than the sum of incremental irreversible social benefits and incremental reversible social net benefits for introducing a GM banana, such that

$$I < I^* = \frac{W}{\beta/(\beta - 1)} + R. \tag{4A.1}$$

The estimation of the MISTICs (I^*) requires quantification of three factors: SIRBs from GM crops (W); the social incremental irreversible benefits (R) rate; and the hurdle rate, $\beta/(\beta - 1)$. All these factors can be estimated or calculated using econometric and mathematical modeling techniques following Demont, Wesseler, and Tollens (2004).

We computed the SIRBs at time t (SIRB(t)) as the SIRBs at complete adoption multiplied by the adoption rate at time t ($\rho(t)$) and multiplied by the exponential factor of the expected growth (or drift) at rate α: $SIRB(t) \cdot \rho(t) \cdot e^{\alpha t}$. The discounted sum of SIRBs ($SIRB_{PV}$) for Uganda over time is calculated as

$$SIRB_{PV} = \int_0^{\cdot} SIRB(t)e^{-(\mu-\alpha)t}dt, \tag{4A.2}$$

where μ is the risk-adjusted discount rate, and α is the drift rate of the geometric Brownian motion, explained in more detail in Kikulwe, Wesseler, and Falck-Zepeda (2008).

We also tried to identify the social incremental irreversible benefits on a per hectare basis using information provided by Bagamba (2007). Most banana producers in Uganda do not use pesticides or fungicides to manage pests and diseases, as mentioned earlier. A small proportion (less than a quarter) of banana producers applies small amounts of pesticides.

The different hurdle rates, $\beta/(\beta - 1)$, were calculated defining β as follows (see Dixit and Pindyck 1994, 147–152):

$$\beta = \frac{1}{2} - \frac{r - \delta}{\sigma^2} + \sqrt{\left[\frac{r-\delta}{\sigma^2} - \frac{1}{2}\right]^2 + \frac{2r}{\sigma^2}} > 1, \tag{4A.3}$$

where r is the risk-free rate of return; δ is the convenience yield defined as the difference between the risk-adjusted discount rate μ and the drift rate α (that is, $\delta = \mu - \alpha > 0$, $\mu \geq r$); and α and σ^2 (variance rate) as before. The maximum likelihood estimators for α and σ^2 were estimated following Campbell, Lo, and MacKinlay (1997).

In our analysis, we limit ourselves to the private incremental reversible benefits at the farm level, assuming all the rents from the new technology are captured by farmers. In the longer run, the rents will be distributed among farmers, the agents in the banana supply chain, and banana consumers. Additional secondary benefits, such as improved food security and reduced vulnerability to external shocks, may be generated through higher farm income among banana growers. Assessing such benefits would require the use of a general equilibrium model for Uganda and are beyond the scope of this study. Thus, the computed SIRBs are equal to the private incremental reversible benefits.

APPENDIX 4B: The Random Parameter Model

An RPLM, or a mixed logit model, is a model that accounts for preference heterogeneity by using a random parameter component to the vector of coefficients (βs). The RPLM does not require the independence-of-irrelevant-alternatives (IIA) assumption. It can also account for unobserved, unconditional heterogeneity in preferences across respondents, even when conditional heterogeneity has been considered, as well as correlation among choices arising from the repetition of choices by the same respondent (McFadden and Train 2000; Garrod, Scarpa, and Willis 2002). The random utility function in the RPLM is given by

$$U_{ij} = V(Z_j(\beta + \eta_i)) + e(Z_j). \tag{4B.1}$$

The utility is decomposed into a deterministic component V and an error component stochastic term e. Indirect utility is assumed to be a function of the choice attributes Z_j, with the utility parameter vector β, which due to preference heterogeneity may vary across respondents by a random component η_i. By specifying the distribution of the error terms e and η_i, the probability of choosing j in each of the choice sets can be derived (Train 1998). By accounting for unobserved heterogeneity, the random parameter logit model takes the form

$$P_{ij} = \frac{\exp(V(Z_j(\beta + \eta_i)))}{\sum_{h=1}^{C}\exp(V(Z_h(\beta = \eta_i)))}. \tag{4B.2}$$

Because this model is not restricted by the IIA assumption, the stochastic part of utility may be correlated among alternatives and across the sequence of choices using the common influence of η_i. Treating preference parameters as random variables requires estimation by simulated maximum likelihood. The maximum likelihood algorithm searches for a solution by simulating k draws from distributions with given means and standard deviations. Probabilities are calculated by integrating the joint simulated distribution.

Even though unobserved heterogeneity can be accounted for in the RPLM, this model fails to explain the sources of heterogeneity (Boxall and Adamowicz 2002). One solution to detecting the sources of heterogeneity while accounting for unobserved heterogeneity could be to include interactions of respondent-specific household characteristics with choice-specific attributes in the utility function. The RPLM with interactions can detect preference variation in terms of the unconditional heterogeneity of tastes (random heterogeneity) and individual characteristics (conditional heterogeneity), so improving the fit of the model (Revelt and Train 1998; Morey and Rossmann 2003).

When the interaction terms are included in the utility function, the indirect utility function that is estimated becomes (Rolfe, Bennett, and Louviere 2000)

$$V_{ij} = \beta + \beta_1 Z_1 + \beta_2 Z_2 + \ldots + \beta_n Z_n + \delta_1(Z_1 S_1) + \delta_2(Z_2 S_2) + \ldots + \delta_l(Z_n S_m). \quad (4\text{B}.3)$$

In this specification, m is the number of respondent-specific characteristics that explain the choice of a banana bunch, and δ_1 to δ_2 is the l-dimensional matrix of coefficients corresponding to the vector of interaction terms S that influence utility. Because respondent-specific characteristics are constant across choice occasions for any given respondent, respondent characteristics only enter as interaction terms with the banana bunch attributes.

Empirically, equation 4B.3 was then extended to include the 42 interactions between the six banana bunch attributes and the seven respondent-specific characteristics:

$$\begin{aligned}
V_{ij} = {} & \beta + \beta_1(Z_{BUNMED}) + \beta_2(Z_{BUNLAR}) + \beta_3(Z_{BENMED}) + \beta_4(Z_{BENLAR}) \\
& + \beta_5(Z_{GMTEC}) + \beta_6(Z_{PRICE}) + \delta_1(Z_{BUNMED} \times S_{HHSIZE}) \\
& + \delta_2(Z_{BUNLAR} \times S_{HHSIZE}) + \ldots + \delta_{42}(Z_{PRICE} \times S_{SWEST}),
\end{aligned} \quad (4\text{B}.3')$$

where β refers to the alternative specific constant (ASC), which was set equal to 1 if either option A or B was chosen and 0 if the respondent chooses the status quo (option C) (Louviere et al. 2000),[9] Z_{BUNMED} is the medium bunch size, Z_{BUNLAR} the large bunch size, Z_{BENMED} the medium benefit, Z_{BENLAR} the large benefit, Z_{GMTEC} GM biotechnology, and Z_{PRICE} the percentage price change. Based on the correlation matrices and variance inflation factors (VIF)[10] results, seven consumer characteristics were retained and interacted with the five banana attributes levels to investigate the possible sources of heterogeneity. The selected consumer characteristics included: (1) household size ($HHSIZE$); (2) whether or not the respondent had postsecondary education ($EDUC$); (3) log of household monthly income ($INCOME$); (4) age of the respondent (AGE); (5) whether or not the household grows bananas ($GROW$); (6) whether or not the household was found in the Eastern region ($EAST$); and (7) whether or not the household was found in the Southwestern region ($SWEST$).

APPENDIX 4C: The Latent Class Model

The latent class model (LCM) casts heterogeneity as a discrete distribution by using a specification based on the concept of endogenous (or latent) preference segmentation (Wedel and Kamakura 2000). The approach describes a population as consisting of a finite and identifiable number of groups of individuals called segments. Preferences are relatively homogeneous within segments but differ substantially across segments. The number of segments is determined endogenously by the data. The insertion of an individual into a specific segment is probabilistic and depends on the characteristics of the respondents. In the model, respondent characteristics indirectly affect the choices through their impact on segment membership.

In the LCM used here (see Kikulwe et al. 2011), the utility that consumer i, who belongs to a particular segment s, derives from choosing banana bunch alternative $j \in C$ can be written as

$$U_{ij/s} = \beta_s X_{ij} + \varepsilon_{ij/s}, \qquad (4C.1)$$

---

9   A fairly more negative and significant ASC indicates a higher tendency of the respondent to choose the status quo.

10   VIF for each regression is calculated as $VIF = \dfrac{1}{1 - R^2}$, where R^2 is the R^2 of the artificial regression with the ith independent variable as a "dependent" variable. Independent variables which exhibited VIF > 5 are eliminated, indicating that they are affected by multicollinearity (Maddala 2001).

where X_{ij} is a vector of attributes associated with banana bunch alternative j of a choice set C and consumer i, and β_s is a segment-specific vector of taste parameters. The differences in β_s vectors enable this approach to capture the heterogeneity in banana bunch attribute preferences across segments. Assuming that the error terms are identically and independently distributed (IID) and follow a Type I distribution, the probability $P_{ij/s}$ of alternative j being chosen by the ith individual in segment s is then given by

$$P_{ij/s} = \frac{\exp(\beta_s X_{ij})}{\sum_{h=1}^{C}\exp(\beta_s X_{ih})}. \tag{4C.2}$$

A membership likelihood function M^* is introduced to classify the consumer into one of the S finite number of latent segments with some probability, P_{is}. The membership likelihood function for consumer i and segment s is given by $M_{is}^* = \lambda_s Z_i + \xi_{is}$, where Z represents the observed characteristics of the household, $\lambda_k (k = 1, 2, \ldots, S)$ is the segment-specific parameters to be estimated, and ξ_{is} is the error term. Assuming that the error terms in the consumer membership likelihood function are IID across consumers and segments and follow a Type 1 distribution, the probability that consumer i belongs to segment s can be expressed as

$$P_{is} = \frac{\exp(\lambda_s X_i)}{\sum_{k=1}^{S}\exp(\lambda_k X_i)}. \tag{4C.3}$$

The segment-specific parameters λ_k denote the contributions of the various consumer characteristics to the probability of segment membership, P_{is}. A positive (negative) and significant λ implies that the associated consumer characteristic, Z_i, increases (decreases) the probability that the consumer i belongs to segment s. P_{is} sums to one across the S latent segments, where $0 \leq P_{is} \leq 1$.

By bringing equations 4C.2 and 4C.3 together, we can construct a mixed-logit model that simultaneously accounts for banana bunch choice and segment membership. The joint unconditional probability of individual i belonging to segment s and choosing banana bunch alternative j can be given by

$$P_{ijs} = (P_{ij/s}) \times (P_{is}) = \left[\frac{\exp(\beta_s X_{ij})}{\sum_{h=1}^{C}\exp(\beta_s X_{ih})}\right]\left[\frac{\exp(\lambda_s X_i)}{\sum_{k=1}^{S}\exp(\lambda_k X_i)}\right]. \tag{4C.4}$$

The two models (appendixes 4B and 4C) used have advantages and disadvantages. The RPLM and LCM, though they both focus on the deterministic component of utility, capture heterogeneity differently. The RPLM captures heterogeneity at the individual level, but it assumes the distribution of taste preferences across the population. The LCM is less flexible (that is, the attribute and variable parameters in each segment are fixed), but it allows explaining preference heterogeneity across segments of a given population. In our data set we used both approaches, but we later applied statistical methods to choose which approach fits our data best (for model comparison, see Colombo, Hanlay, and Louviere 2009).

APPENDIX 4D: Compensating Surplus Welfare Analysis

The best-fit LCM,[11] which is used to group the population into homogeneous segments, was employed to estimate the required parameters for welfare measures. The LCM allows us to calculate WTP welfare measures for each respondent in a segment. Deriving welfare measures under the LCM is done in two steps. First, policy impacts at the segment level are identified by calculating WTP welfare measures for each segment. Second, the standard aggregate procedure that assumes homogeneous preference is corrected for heterogeneity. That is done by computing the weighted sum of segment-specific welfare measures. The weights are the estimated individual segment membership probabilities (Boxall and Adamowicz 2002). The individual segment WTP can finally be aggregated to calculate WTP welfare measures for the whole population.

The compensating surplus (*CS*) welfare measure for changes in the banana bunch attributes, conditional on the segment membership, can be derived from the estimated parameters by using the following equation (Bateman et al. 2003):

$$CS_n = \frac{1}{\Theta_{price}} \left[\ln \sum_{i \in C} e^{V_i^1} - \ln \sum_{i \in C} e^{V_i^0} \right], \tag{4D.1}$$

---

11   Following Colombo, Hanlay, and Louviere (2009), a comparison between RPLM with interactions (rural sample) and LCM results into a probability of $P \leq \Phi(-37.20) \cong 0$ and that for RPLM with interactions (urban sample) and LCM gives $P \leq \Phi(-26.55) \cong 0$. That is, for RPLM with interactions for urban ($K_2 = 16$) and LCM ($K_1 = 19$), $\rho_2^2 - \rho_1^2 = 0.198$, while for RPLM with interactions for rural ($K_2 = 18$) and LCM, $\rho_2^2 - \rho_1^2 = 0.207$. This indicates that the null hypothesis is rejected; hence the LCM is preferred. These results show that the preference heterogeneity in our data is better accounted for at the segment level rather than at the individual level.

where the compensating surplus CS_n is the amount of money that one would have to give the individual n after the change has occurred for that person to remain as well off as before (that is, after choosing alternative i in the choice set C); Θ_{price} is the marginal utility of money and is the coefficient of the banana bunch price attribute; V_i^0 represents the individual's utility at the initial level (that is, the current state: banana bunches bred through traditional biotechnology); and V_i^1 is the utility of the alternative level (that is, after change state: bunches bred by GM biotechnology) following changes in attributes.

The final marginal WTP welfare measure can be derived by first integrating the welfare effects across the different segments,

$$CS_{n|s} = \frac{1}{\Theta_{price_s}}\left[\ln\sum_{i\in C}e^{V_i^1} - \ln\sum_{i\in C}e^{V_i^0}\right], \qquad (4D.2)$$

where $s = 1, 2$ is the number of segments, and Θ_{price} is the coefficient on the banana bunch price attribute for each segment providing each segment's marginal utility of income; and then by calculating the weighted sum of the segment membership:

$$CS_{n|s}^t = \sum_{i\in C}^{2}W_{ns}\left\{\frac{1}{\Theta_{price_s}}\left[\ln\sum_{i\in C}e^{V_i^1} - \ln\sum_{i\in C}e^{V_i^0}\right]\right\}, \qquad (4D.3)$$

where W_{ns} is the probability of an individual n being in segment s.

To estimate the consumers' compensating surplus (CS), conditional on being in segment 1 or 2, for introduction of a GM banana over the status quo, four possible options were created. The creation of the four policy-relevant scenarios was based on the banana bunch profiles. The attribute levels that characterize a number of alternative banana bunch improvements scenarios are listed below, along with the base case:

- *Base case* (small bunches with no benefits)—status quo: this is the baseline situation where banana bunches consumed are mostly of small bunch sizes, produced through traditional biotechnology. The price for the base case is at UGX 3000 for a 10-kilogram bunch.

- *Scenario 1* (all medium improvement): medium banana bunch size produced by GM biotechnology that generates medium benefits (in the form of increased yields) for producers.

- *Scenario 2* (all large improvement): large banana bunch size produced by GM biotechnology that generates large benefits for producers.

- *Scenario 3* (large bunch with medium benefits): large banana bunch size produced by GM biotechnology that generates medium benefits for producers.

- *Scenario 4* (medium bunch with large benefits): medium banana bunch size produced by GM biotechnology that generates large benefits for producers.

To find the *CS* associated with each of the above scenarios, the difference between the welfare measures under status quo and the four banana bunch options was calculated.

For the consumer segment that was found to have negative WTP (the potential GM banana opponents), the next step was to find out whether their negative WTP was below or above the MISTICs. We compared the MISTICs per bunch with the estimated total WTP per bunch. To calculate the MISTICs per bunch, we divided the annual MISTICs per household by the average number of bunches consumed per household. Using a per capita consumption of cooking bananas in Uganda of 250 kg per year (NARO 2001), an average bunch size of 10 kilograms (baseline), and an average household size of the potential GM banana opponents of 5.67 members per household (estimated under research question 4; see Kikulwe et al. 2011), and dividing the annual consumption with the average bunch size resulted in an average per capita consumption of 25 bunches per year. The product of the per capita bunches consumed and household size yielded an average of approximately 142 bunches consumed per household per year. Dividing the MISTICs per household by the number of bunches consumed per household and year after deducting the planting costs provided the MISTICs per bunch.

References

ASTI (Agricultural Science and Technology Indicators). 2002. Country Brief 1. Washington, DC: International Food Policy Research Institute. www.asti.cgiar.org/uganda.

Atanassov, A., A. Bahieldin, J. Brink, M. Burachik, J. I. Cohen, V. Dhawan, R. V. Ebora, et al. 2004. *To Reach the Poor: Results from the ISNAR-IFPRI Next Harvest Study on Genetically Modified Crops, Public Research, and Policy Implications.* IFPRI Discussion Paper 116. Washington, DC: International Food Policy Research Institute.

Bagamba, F. 2007. "Market Access and Agricultural Production: The Case of Banana Production in Uganda." PhD Thesis, Wageningen University, Wageningen, the Netherlands.

Bateman, I. J., R. T. Carson, B. Day, W. M. Hanemann, N. Hanley, T. Hett, M. Jones-Lee, et al. 2003. *Guidelines for the Use of Stated Preference Techniques for the Valuation of Preferences for Non-market Goods.* Cheltenham, UK: Edward Elgar.

Beckmann, V., C. Soregaroli, and J. Wesseler. 2006a. *Governing the Co-existence of GM crops: ex ante Regulation and ex post Liability under Uncertainty and Irreversibility.* Institutional Change in Agriculture and Natural Resources Discussion Paper 12. Berlin: Humboldt University.

———. 2006b. "Co-existence Rules and Regulations in the European Union." *American Journal of Agricultural Economics* 88 (5): 1193–1199.

Birol, E., F. R. Villaba, and M. Smale. 2009. "Farmer Preferences for *Milpa* Diversity and Genetically Modified Maize in Mexico: A Latent Class Approach." *Environment and Development Economics* 14 (4): 521–540.

Boxall, P. C., and W. L. Adamowicz. 2002. "Understanding Heterogeneous Preferences in Random Utility Models: A Latent Class Approach." *Environmental and Resource Economics* 23: 421–446.

Breffle, W., and E. Morey. 2000. "Investigating Preference Heterogeneity in a Repeated Discrete-Choice Recreation Demand Model of Atlantic Salmon Fishing." *Marine Resource Economics* 15: 1–20.

Brenner, C. 2004. *Telling Transgenic Technology Tales: Lessons from the Agricultural Biotechnology Support Project (ABSP) Experience.* ISAAA Brief 31. Ithaca, NY, US: International Service for the Acquisition of Agri-Biotech Applications.

Campbell, J. Y., A. W. Lo, and A. C. MacKinlay. 1997. *The Econometrics of Financial Markets.* Princeton, NJ, US: Princeton University Press.

Carlsson, F., P. Frykblom, and C. Liljenstolpe. 2003. "Valuing Wetland Attributes: An Application of Choice Experiments." *Ecological Economics* 47: 95–103.

Clarke, T. 2003. "Banana Lab Opens in Uganda: Genetic Modification of Clonal Crop Could Soon Follow." *Nature News,* August 22. www.bioedonline.org/news/news.cfm?art=430.

Colombo, S., N. Hanlay, and J. Louviere. 2009. "Modeling Preference Heterogeneity in Stated Choice Data: An Analysis for Public Goods Generated by Agriculture." *Agricultural Economics* 40 (3): 307–322.

Cox, D., and H. Miller. 1965. *The Theory of Stochastic Processes.* London: Chapman and Hall.

Demont, M., J. Wesseler, and E. Tollens. 2004. "Biodiversity versus Transgenic Sugar Beet: The One Euro Question." *European Review of Agricultural Economics* 31 (1): 1–18.

Dixit, A., and R. S. Pindyck. 1994. *Investment under Uncertainty.* Princeton, NJ, US: Princeton University Press.

Edmeades, S. 2003. "Variety Choice and Attribute Trade-offs within the Framework of Agricultural Household Models: The Case of Bananas in Uganda." PhD Thesis, North Carolina State University, Raleigh, NC, US.

Edmeades, S., and M. Smale. 2006. "A Trait-Based Model of the Potential Demand for a Genetically Engineered Food Crop in a Developing Economy." *Agricultural Economics* 35: 351–361.

Falck-Zepeda, J. B. 2006. "Coexistence, Genetically Modified Biotechnologies, and Biosafety: Implications for Developing Countries." *American Journal of Agricultural Economics* 88 (5): 1200–1208.

FAO (Food and Agriculture Organization of the United Nations). 2006. *FAOSTAT.* Accessed February 13, 2006. http://faostat.fao.org/site/567/default.aspx#ancor.

Garrod, G. D., R. Scarpa, and K. G. Willis. 2002. "Estimating the Benefits of Traffic Calming on through Routes: A Choice Experiment Approach." *Journal of Transport Economics and Policy* 36 (2): 211–231.

Gaskell, G., A. Allansdottir, N. Allum, C. Corchero, C. Fischler, J. Hampel, J. Jackson, et al. 2006. *Europeans and Biotechnology in 2005: Patterns and Trends.* Eurobarometer 64.3. Brussels: European Commission.

Gold, C. S. 1998. "Banana Weevil: Ecology Pest Status and Prospects for Integrated Control with Emphasis on East Africa." In *Proceedings of a Symposium on Biological Control in Tropical Habitats: Third International Conference on Tropical Entomology,* edited by S. K. Saini, 49–74. Nairobi, Kenya: International Centre of Insect Physiology and Ecology.

———. 2000. "Biology and Integrated Pest Management of Banana Weevil, *Cosmopolites sordidus* (Germar)." In *Advancing Banana and Plantain R&D in Asia and the Pacific,* vol. 10, edited by A. B. Molina, V. N. Roa, and M.A.G. Maghuyop, 28–33. Los Baños, Laguna, Philippines: International Network for the Improvement of Banana and Plantain—Asia and the Pacific Network.

Gold, C. S., J. E. Pena, and E. B. Karamura. 2001. "Biology and Integrated Pest Management for the Banana Weevil, *Cosmopolites sordidus* (Germar) (*Coleoptera: Curculionidae*)." *Integrated Pest Management Reviews* 6 (2): 79–155.

Gold, C. S., G. Night, A. Abera, and P. R. Speijer. 1998. "Hot-Water Treatment for the Control of the Banana Weevil, *Cosmopolites sordidus* Germar (*Coleoptera: Curculionidae*), in Uganda." *African Entomology* 6 (2): 215–221.

GOU (Government of Uganda). 2002a. *Guidelines on Biosafety in Biotechnology for Uganda.* Kampala, Uganda: Uganda National Council of Science and Technology.

———. 2002b. *National Biodiversity Strategy and Action Plan: National Environment Management Authority (NEMA).* Accessed June 1, 2006. www.biodiv.org/doc/world/ug/ug-nbsap-01-en .doc.

————. 2004. *Draft of National Biotechnology and Biosafety Policy.* Kampala, Uganda: Uganda National Council of Science and Technology.

————. 2005. *The Uganda Biosafety Bill.* Kampala, Uganda.

————. 2009. "Draft Estimates of Revenue and Expenditure (Recurrent and Development). FY2009/10." Accessed September 24, 2009. www.finance.go.ug/docs/Draft%20Budget %20Estimates%20FY2009_10%207July%2009_with%20page%20numbers.pdf.

————. 2012. "The Background to the Budget 2012/13 Fiscal Year: Priorities for Renewed Economic Growth and Development." Accessed January 28, 2013. www.finance.go.ug/index .php?option=com_docman&Itemid=117.

Hu, W., A. Hünnemeyer, M. Veeman, W. L. Adamowicz, and L. Srivastava. 2004. "Trading off Health, Environmental and Genetic Modification Attributes in Food." *European Review of Agricultural Economics* 31: 389–408.

Jaffe, G. 2006. *Comparative Analysis of the National Biosafety Regulatory Systems in East Africa.* IFPRI Discussion Paper 146. Washington, DC: International Food Policy Research Institute.

Just, R. E., D. L. Hueth, and A. Schmitz. 2004. *The Welfare Economics of Public Policy: A Practical Approach to Project and Policy Evaluation.* Cheltenham, UK: Edward Elgar.

Kalaitzandonakes, N., J. Alston, and K. Bradford. 2007. "Compliance Costs for Regulatory Approval of New Biotech Crops." *Nature Biotechnology* 25 (5): 509–511.

Karamura, D. A. 1998. "Numerical Taxonomic Studies of the East African Highland Bananas (Musa AAA–East Africa) in Uganda." PhD Thesis, University of Reading, UK.

Kikulwe, E. M. 2010. "On the Introduction of Genetically Modified Bananas in Uganda: Social Benefits, Costs, and Consumer Preferences." PhD Thesis, Wageningen University, the Netherlands.

Kikulwe, E. M., J. Wesseler, and J. Falck-Zepeda. 2008. *Introducing a Genetically Modified Banana in Uganda: Social Benefits, Costs, and Consumer Perceptions.* IFPRI Discussion Paper 767. Washington, DC: International Food Policy Research Institute.

Kikulwe, E. M., K. Nowakunda, M.S.R. Byabachwezi, J. M. Nkuba, J. Namaganda, D. Talengera, E. Katungi, and W. K. Tushemereirwe. 2007. "Development and Dissemination of Improved Banana Cultivars and Management Practices in Uganda and Tanzania." In *An Economic Assessment of Banana Genetic Improvement and Innovation in the Lake Victoria Region of Uganda and Tanzania,* edited by M. Smale and W. K. Tushemereirwe. Research Report 155, 37–48. Washington, DC: International Food Policy Research Institute.

Kikulwe, E. M., E. Birol, J. Wesseler, and J. Falck-Zepeda. 2011. "A Latent Class Approach to Investigating Demand for Genetically Modified Banana in Uganda." *Agricultural Economics* 42 (5): 547–560.

Knight, J. A., D. W. Mather, D. K. Holdsworth, and D. F. Ermen. 2007. "Acceptances of GM Food—An Experiment in Six Countries." *Nature Biotechnology* 25 (5): 507–508.

Kontoleon, A. 2003. "Essays on Non-market Valuation of Environmental Resources: Policy and Technical Explorations." PhD Thesis, University College London.

Kontoleon, A., and M. Yabe. 2006. "Market Segmentation Analysis of Preferences for GM Derived Animal Foods in the UK." *Journal of Agricultural and Food Industrial Organization* 4 (1): 8. Accessed July 15, 2007. www.bepress.com/jafio/vol4/iss1/art8.

Lancaster, K. 1966. "A New Approach to Consumer Theory." *Journal of Political Economy* 84: 132–157.

Li, Q., K. R. Curtis, J. J. McCluskey, and T. I. Wahl. 2003. "Consumer Attitudes towards Genetically Modified Foods in Beijing, China." *AgBioForum* 5 (4): 145–152.

List, J., and C. Gallet. 2001. "What Experimental Protocols Influence Disparities between Actual and Hypothetical Stated Values?" *Environmental and Resource Economics* 20 (3): 241–254.

List, J., P. Sinha, and M. Taylor. 2006. "Using Choice Experiments to Value Non-market Goods and Services: Evidence from Field Experiments." *Advances in Economic Analysis and Policy* 6 (2): 1–37.

Loureiro, M. L., and M. Bugbee. 2005. "Enhanced GM Foods: Are Consumers Ready to Pay for the Potential Benefits of Biotechnology?" *Journal of Consumer Affairs* 39 (1): 52–70.

Louviere, J. J., D. A. Hensher, J. D. Swait, and W. L. Adamowicz. 2000. *Stated Choice Methods: Analysis and Applications*. Cambridge: Cambridge University Press.

Maddala, G. S. 2001. *Introduction to Econometrics,* 3rd ed. Saddle River, NJ, US: Prentice-Hall.

McFadden, D., and K. Train. 2000. "Mixed MNL Models of Discrete Response." *Journal of Applied Economics* 56: 162–175.

Mitthöfer, D. 2005. "Economics of Indigenous Fruit Tree Crops in Zimbabwe." PhD Thesis, Hannover University, Hannover, Germany.

Modgil, M., K. Mahajan, S. K. Chakrabarti, D. R. Sharma, and R. C. Sobti. 2005. "Molecular Analysis of Genetic Stability in Micropropagated Apple Rootstock MM106." *Scientia Horticulturae* 104: 151–160.

Morey, E., and K. Rossmann. 2003. "Using Stated-Preference Questions to Investigate Variations in Willingness to Pay for Preserving Marble Monuments: Classic Heterogeneity, Random Parameters, and Mixture Models." *Journal of Cultural Economics* 27: 215–229.

Nampala, P., C. Mugoya, and T. Ssengooba. 2005. *Biosafety Regulatory System in Uganda*. Presentation to the Program for Biosafety Systems Roundtable, April 18–20, in Entebbe, Uganda.

NARO (National Agricultural Research Organisation). 2001. *Multilocational Banana Germplasm Evaluation Trials*. Third report. Entebbe, Uganda.

Nowakunda, K. 2001. "Determination of Consumer Acceptability of Introduced Bananas." Masters Thesis, Makerere University, Kampala, Uganda.

Nowakunda, K., P. R. Rubaihayo, M. A. Ameny, and W. K. Tushemereirwe. 2000. "Acceptability of Introduced Bananas in Uganda." *InfoMusa* 9 (2): 22–25.

OPM (Oxford Policy Management). 2005. *Evaluation of the Plan for the Modernization of Agriculture.* Main report. Oxford. www.pma.go.ug/pmauploads/Main%20 PMA %20evaluation%20final%20report.pdf.

Ouma, E., A. Abdulai, and A. Drucker. 2007. "Measuring Heterogeneous References for Cattle Traits among Cattle-Keeping Households in East Africa." *American Journal of Agricultural Economics* 89: 1005–1019.

Owen, K., J. Louviere, and J. Clark. 2005. *Impact of Genetic Engineering on Consumer Demand.* Final report. RIRDC Publication 05/015. Kingston, Australia: Rural Industries Research and Development Corporation. www.rirdc.gov.au/fullreports/index.htm.

Paarlberg, R. 2008. *Starved for Science: How Biotechnology Is Being Kept out of Africa.* Cambridge, MA, US: Harvard University Press.

Revelt, D., and K. Train. 1998. "Mixed Logit with Repeated Choices: Households' Choice of Appliance Efficiency Level." *Review of Economics and Statistics* 53: 647–657.

Rolfe, J., J. Bennett, and J. Louviere. 2000. "Choice Modelling and Its Potential Application to Tropical Rainforest Preservation." *Ecological Economics* 35 (2): 289–302.

Rubaihayo, P. R. 1991. *Banana-Based Cropping Systems Research: A Report on Rapid Appraisal Survey of Banana Production in Uganda.* Research Bulletin 1. Kampala, Uganda: Department of Crop Science, Makerere University.

Rubaihayo, P. R., and C. Gold. 1993. *Banana-Based Cropping Systems Research: A Report on Rapid Appraisal Survey of Banana Production in Uganda.* Research Bulletin 2. Kampala, Uganda: Department of Crop Science, Makerere University.

Ruto, E., G. Garrod, and R. Scarpa. 2008. "Valuing Animal Genetic Resources: A Choice Modeling Application to Indigenous Cattle in Kenya." *Agricultural Economics* 38: 89–98.

Scarpa, R., A. Drucker, S. Anderson, N. Ferraes-Ehuan, V. Gomez, C. R. Risopatron, and O. Rubio-Leonel. 2003. "Valuing Animal Genetic Resources in Peasant Economies: The Case of the Box Keken Creole Pig in Yucatan." *Ecological Economics* 45 (3): 427–443.

Scatasta, S., J. Wesseler, and M. Demont. 2006. "Irreversibility, Uncertainty, and the Adoption of Transgenic Crops: Experiences from Applications to Ht Sugar Beet, Ht Corn, and Bt Corn." In *Regulating Agricultural Biotechnology: Economics and Policy,* edited by R. E. Just, J. M. Alston, and D. Zilberman, 327–352. New York: Springer.

Shotkoski, F. A., L. Tripathi, A. Kiggundu, G. Arinaitwe, and W. K. Tushemereirwe. 2010. "Role of Biotechnology and Transgenics in Bananas (*Musa* spp.) in Africa." In *International Conference on Banana and Plantain in Africa: Harnessing International Partnerships to Increase Research Impact,* edited by T. Dubois, S. Hauser, C. Staver, and D. Coyne, 275–279. *Acta Horticulturae (International Society for Horticultural Science)* 879: 275–279.

Smale, M., and W. K. Tushemereirwe, eds. 2007. *An Economic Assessment of Banana Genetic Improvement and Innovation in the Lake Victoria Region of Uganda and Tanzania.* Research Report 155. Washington, DC: International Food Policy Research Institute.

Ssebuliba, R. N. 2001. "Fertility in East African Highland Bananas." PhD Thesis, Makerere University, Kampala, Uganda.

Ssebuliba, R. N., D. Talengera, D. Makumbi, P. Namanya, A. Tenkouano, W. K. Tushemereirwe, and M. Pillay. 2006. "Reproductive Efficiency and Breeding Potential of East African Highland (Musa AAA–EA) Bananas." *Field Crops Research* 95: 250–255.

Stover, R. H., and N. W. Simmonds. 1987. *Bananas,* 3rd ed. London: Longman.

Swennen, R., and D. R. Vuylsteke. 1991. "Bananas in Africa: Diversity, Uses and Prospects for Improvement." In *Crop Genetic Resources for Africa,* edited by N. Q. Ng, P. Perrino, F. Attere, and H. Zedah, vol. 2, 151–160. London: Trinity Press.

Train, K. E. 1998. "Recreation Demand Models with Taste Differences over People." *Land Economics* 74: 230–239.

Traynor, P. 2003. "Uganda Biotechnology Project." Final draft report submitted to the US Agency for International Development, RAISE Task Order 817, Washington, DC.

Tripathi, L. 2003. "Genetic Engineering for Improvement of *Musa* Production in Africa." *African Journal of Biotechnology* 2 (12): 503–508.

Tushemereirwe, W. K., A. Kangire, J. Smith, F. Ssekiwoko, M. Nakyanzi, D. Kataama, C. Musiitwa, and R. Karyeija. 2003a. "An Outbreak of Banana Bacterial Wilt on Banana in Uganda." *InfoMusa* 12 (2): 6–8.

Tushemereirwe, W. K., I. Kashaija, W. Tinzaara, C. Nankinga, and S. New. 2003b. *Banana Production Manual: A Guide to Successful Banana Production in Uganda,* 2nd ed. Kampala, Uganda: Makerere University Printery.

UBOS (Uganda Bureau of Statistics). 2006. *National Statistics Databank, Agriculture and Fisheries. Area Planted and Production of Selected Food Crops, 1980–2004.* Accessed June 15. www.ubos .org.

UNCST (Uganda National Council for Science and Technology). 2006. *A Gap Analysis Study of the Communication and Outreach Strategy for Biotechnology and Biosafety in Uganda.* Kampala, Uganda.

Wafula, D., and N. Clark. 2005. "Science and Governance of Modern Biotechnology in Sub-Saharan Africa—The Case of Uganda." *Journal of International Development* 17: 679–694.

Wedel, M., and W. Kamakura. 2000. *Market Segmentation: Conceptual and Methodological Foundations*. Boston: Kluwer Academic.

Wesseler, J., S. Scatasta, and E. Nillesen. 2007. "The Maximum Incremental Social Tolerable Irreversible Costs (MISTICs) and Other Benefits and Costs of Introducing Transgenic Maize in the EU-15." *Pedobiologia* 51 (3): 261–269.

Chapter 5

Genetically Modified Organisms, Exports, and Regional Integration in Africa

David Wafula and Guillaume Gruère

As noted in the introduction to this volume, genetically modified (GM) crops have been characterized by high adoption rates and a steady increase in global market value. However, their rapid diffusion has triggered a diversity of concerns, including issues related to the safety of the technology to human health and the environment, socioeconomic considerations, and issues of ownership and control. Trade-related impacts and access to export markets are increasingly emerging as another major concern. Destinations such as the European Union (EU), where the level of caution and consumer scepticism is relatively high, have attracted a lot of attention (see, for example, Gruère 2006; Paarlberg 2009).

More specifically, several African countries have been preoccupied with the notion that adoption of GM crops would attract a blanket rejection of agricultural exports by importing countries in the Western world, especially by European countries (see, for example, Gruère and Sengupta 2009; Paarlberg 2009). The dilemma in policymaking circles has been how to harness the potential benefits of GM crops while preserving trade interests and niche markets (Anderson and Jackson 2005). Although some countries have taken precautionary stances, others have decided to take a "wait and see" stance (Gruère 2006). The potential economic risk for these countries of taking conservative GM-free policy positions is that they may deny farmers the opportunity to harness and maximize the potential benefits of the technology. By focusing on avoiding any potential future export risks with Europe or other developed nations, these countries may have filtered out nonrisky but beneficial technology.

At the same time, little attention has been accorded to the intraregional trade dimension, which is increasingly becoming fundamental (Diao et al. 2005). Intraregional trade in future commercialized products is going to be an issue that may affect trade and the entire regional integration efforts unless biosafety regulatory mechanisms are put in place to address them.

This chapter reviews results from the literature to discuss issues of GM product use, export risk management, and regional integration in Africa. The next section focuses on possible export risks associated with the use of GM crops outside of the region, and the subsequent section discusses the increasing role of regional harmonization schemes. The chapter concludes with some policy lessons.

Implications for Exports of Commercializing GM Crops

Agriculture is the economic pillar of most African countries south of the Sahara. About 70 percent of the population in most African countries is rurally based and depends on agriculture as a source of income and livelihood. In this regard, efforts related to the promotion of agricultural productivity have been placed at the top of Africa's integration priorities and processes. The agricultural sector in the region is facing many challenges, including production constraints at the farm level. In exploring a range of technological options that can supplement conventional tools, the potential of agricultural biotechnology has been recognized, even though reservations and resistance from some quarters still prevail (Juma and Serageldin 2007). The issue of possible market-access barriers in key export destinations has been an issue of paramount concern. However, the magnitude of the anticipated or perceived risks remains to be analyzed and understood in concrete terms.

Three types of approaches have been used in the literature, all directly or indirectly assessing export risks with the adoption of GM crops in specific countries or regions of Africa: qualitative case study analysis linking actors along the commodity chain, bilateral trade data analysis to assess the possible commercial risk, and quantitative economic simulations with trade models under specific scenarios. We review the main results of these three types of studies in the following sections. At the same time, we also discuss the key limitations of each of these approaches.

Case Study Approach

Although several publications discuss the influence of exports on African decisions (for example, Paarlberg 2009; Gruère and Takeshima 2012), to our knowledge, only one paper uses a case study approach to systematically analyze their nature and implications. Gruère and Sengupta (2009) provide a review of international cases where GM-free private standards set up by supermarkets or other buyers in developed countries have affected biosafety decisions, including commercialization in developing countries.

Because their analysis is qualitative and partially relies on secondary data from news clips and media reports that are not always accurate, the evidence may not be fully substantiated. Furthermore, even when based on primary data, the data do not allow measuring the scope or significance of the issues and their precise roles in influencing decisionmaking. The analysis simply points toward reports of influential links between trading actors and decision makers in the area of biosafety.

Gruère and Sengupta (2009) find 29 cases in which export concerns influenced biosafety decisionmaking in 21 countries. They then classify these cases into three categories: cases where the alleged export risks associated with a specific decision on GM crops are largely unfounded or irrational, cases where the export risks and policy decisions are debatable, and cases where decisions are supported by real commercial risks.

Interestingly, Gruère and Sengupta (2009) note that Africa south of the Sahara is the region where the most cases with irrational risks are being reported. Although 11 of the 29 cases are in the first category, 7 of those were reported from Africa. Of the 13 cases related to Africa, 10 are in the first or second categories.

Several cases relate to cash crops. It is traditionally known that African countries have been leading exporters of cash crop commodities, including tea, coffee, cocoa, pyrethrum, sugar, tobacco, bananas, and a wide range of horticultural products to various destinations around the world. GM varieties for these traditional exports have yet to be developed, and the situation in Africa is unlikely to drive any commercial interest in releasing GM varieties of these crops. This being the case, there are unrealistic and unjustified requirements from some of the export destinations.

For instance, Gruère and Sengupta (2009) report that GM-free certification is required for exports of tea from Kenya to the United Kingdom, even though it is widely known that GM tea has not been developed or commercialized anywhere in the world. The product development pathway for genetically modified organisms (GMOs) shows that it takes about 10 years for a product to pass through the various regulatory steps before it is placed on the market. Development of GMOs also requires a massive capital and research-intensive investment (Sinai 2001).

They also find reports of organic producer groups in Kenya believing that producing GM field crops would jeopardize their exports of horticultural products. The Kenya Organic Association Network fears that introduction of GM varieties of maize or cotton would affect market access for horticultural products that are organically produced. Yet the possibilities of gene-flow contamination from maize or cotton to horticultural products cannot occur, because the products are not biologically compatible (Gruère and Sengupta 2009).

These and other cases provide qualitative evidence of possible irrational decisions in the presence of possible but unproven export risks in African countries. Gruère and Sengupta (2009) attribute this phenomenon to four factors: insufficient knowledge and information on the part of decisionmakers compared to influential agents, risk aversion, and misleading presumptions on the existence of GM open markets and the possibility of segregating non-GM exports. Regardless of the specific reasons, their findings do shine a light on the strict precautionary behavior of African decisionmakers on biosafety and testing or use of GM crops.

Trade Data Analysis

Going one step beyond case studies, simple bilateral trade data analyses have been used to assess the likelihood of immediate export risk with the introduction of a GM crop in a particular country. The principle underlying such studies is that the examination of the composition, destination, and monetary value of commodities exported in past years can help assess the possible short-term trade-related ramifications and market-access barriers that may result from the introduction and commercialization of GM crops in particular countries in Africa.

It should be noted that this approach has a number of critical limitations and does not provide sufficient information to dispel any potential trade risks. Most importantly, it is based on the assumption that past trade data provide a valid representation of future trade relationships; that is, export destinations, trade volume, and value are assumed to remain similar for the period of study and for the near future. Naturally, this is not an innocuous assumption, especially in the case of African countries south of the Sahara, which face a number of productivity and market constraints that may be overcome in the years to come.[1] In particular, if a GM crop was able to increase productivity, it might result in a net marketable surplus that would face export constraints. Furthermore, the assumption disregards the possibility of new trade flows from countries that do not export any potentially GM commodity to Europe and other GM-regulating countries (zero trade). Clearly, the lack of trade today does not mean there is no risk to trade in the future.

---

1 For instance, Juma (2010) argues that Africa has the means to become self-sufficient and even be able to export commodities in the near future.

Still, even accounting for uncertainties, analyzing recent trade data can help discriminate cases that represent immediate or short-run export risks from others based on discourse rather than data. As noted in the previous section, some reported cases have linked seemingly irrational commercial risks to regulatory decisions. Finding that a country has a large trade relationship with a GM-regulating country is useful for managing risks and avoiding immediate export losses. For instance, Thailand and Vietnam, aware of the importance of non-GM rice preferences of major importers, decided to ban any experiment on GM rice to avoid any possible export loss (Gruère and Sengupta 2009). The presence of past trade with large importing countries that regulate or avoid GM food, although not necessarily predicting the future, provides useful information for decisionmakers.[2]

With these caveats in mind, an influential economic analysis was conducted on the trade implications of GMOs in Africa under the aegis of the Common Market for Eastern and Southern Africa (COMESA) (Paarlberg et al. 2006). The Regional Approach to Biotechnology and Biosafety Policy in Eastern and Southern Africa project was initiated to address concerns that transboundary movement of GMOs in the COMESA region might impact trade among member states unless a regional policy mechanism was put in place to mitigate such eventualities. The project covered six case study countries: Egypt, Ethiopia, Kenya, Tanzania, Uganda, and Zambia.

The study analyzed the short-run trade ramifications of introducing and commercializing GMOs that are available globally under recent trade flows to current markets. The magnitude of potential immediate trade losses, based on 2003 bilateral trade data, was illustrated by examining the total value of agricultural commodity exports and the proportion of this export value that risks being rejected in market destinations that treat GM commodities with sensitivity (either due to regulations or buyers' preferences). That GMOs can affect trade cannot be disputed. Consignments of agricultural exports originating from a country that has commercialized GMOs are treated with suspicion and are generally expected to contain GMOs, even in cases where such consignments contain only GM-free products. The Regional Approach to Biotechnology and Biosafety Policy in Eastern and Southern Africa study computed the potential value of conventional agricultural food and feed products exported to various destinations from the six case study countries in 2003. The proportion that is likely to be rejected

---

2 Moreover, given the political realities, short-run considerations are much more powerful in the eyes of decisionmakers than long-term potential issues.

because of GM sensitivity was ascertained and expressed as a percentage of the total agricultural food and feeds exported.

The share of the past exports that are GM sensitive commodities consists of agricultural commodities whose GM counterparts have been approved globally for commercial planting. They include soybeans, maize, cotton, canola, squash, rice, papaya, tomatoes, and Irish potatoes.

The worst-case scenario of the analysis assumed that if the six targeted COMESA countries commercialized the aforementioned products, all exports of food and feed products associated with the crops in question (second column of Table 5.1) would be rejected by all importing destinations worldwide. Naturally, this assumption is exaggerated, given that many countries do not regulate imports of GM products and even those that do—including European countries—import large volumes of approved GM products. The findings revealed that the proportion of exports immediately at risk would be 8.5 percent for Egypt, 2.2 percent for Ethiopia, 1.1 percent for Kenya, 6.2 percent for Tanzania, 6.5 percent for Uganda, and 6.3 percent for Zambia.

Assuming that only the EU would reject the exports, the level of immediate risk for Egypt would be about 4 percent and the rest of the countries 1 percent or less, as shown in Table 5.2 (Paarlberg et al. 2006).

This low level of immediate trading risk exposure stems from the fact that most of the potential GM commodities go to other African countries. Thus, continental and regional demand largely exceeds external demand for the selected products. Although this result does not preclude the possibility that

TABLE 5.1   GM (genetically modified) crops and products included in the Regional Approach to Biotechnology and Biosafety Policy in Eastern and Southern Africa project

Agricultural crops for which GM varieties are approved for commercial planting in at least one country	Possible GM export products
Soybeans	Live animals
Maize	Meat
Cotton	Dairy, eggs, and natural honey
Canola (rape)	Potatoes
Squash	Tomatoes
Papaya	Papaya
Tomatoes	Squash
Irish potatoes	Soybean and rape, including oil, flour, and meal
Sugarbeets	Maize
	Maize flour, meal, and bran
	Maize (hulled)
	Cottonseed, including oil and cake

Source: Paarlberg et al. (2006).

TABLE 5.2   Immediate export losses if all European importers shunned all "possibly GM" or "possibly GM-tainted" products

| Country | Agricultural food and feed-product exports, 2003 | | |
	Total (US$ million)	Of which "possibly GM" and exported to Europe (US$ million)	Share of total exports lost (%)
Egypt	938	37.7	4.0
Ethiopia	450	0.04	0.009
Kenya	1,291	0.03	0.002
Tanzania	408	1.5	0.4
Uganda	116	0.01	0.009
Zambia	119	0.2	0.2

Source: Paarlberg et al. (2006).
Note: GM = genetically modified; US$ = US dollars.

the situation may change, it indicates that the discussed GM products would not face an immediate threat.

Trade Simulations

Several research publications have focused on modeling the economic effects of GM crop adoption in the presence of trade restrictions in developed countries (see Smale et al. 2009). Most of these papers studied the effects of adoption at the global or regional level, and only a few focused on the effects in Africa. Almost all of these papers used a computable general equilibrium model to simulate the effect of adoption or nonadoption of GM crops in specific regions or throughout Africa south of the Sahara.

Although these simulations provide an improved, more detailed, and more structured representation of trade and economic effects, they also face a number of limitations, notably because of their structure and the database they are based on. In particular, they are based on several key assumptions, including the use of known and stable productivity shock and adoption rates in each country of production, masking the variability among producers, production practices (input use), regions, and years; a perfectly competitive world market; and a specific reference year. Their value is to provide some insight into the potential economywide effects of the use of GM crops in the region, based on specified assumptions.

In this category, several studies focused on the effects of adopting GM cotton (for example, Elbehri and MacDonald 2004; Anderson and Valenzuela 2007), a commodity that does not face any trade regulation (ICAC 2010), but whose adoption is found to be necessary in a globally competitive market. Assuming it is productivity enhancing, and knowing that all major countries have adopted GM cotton, there is a significant opportunity cost of nonadoption of GM cotton for countries in the region (Bouët and Gruère 2011).

In the case of GM food and feed crops, Anderson and Jackson (2005) use a global general equilibrium model to simulate scenarios of adoption with a focus on Africa south of the Sahara. Their results show that GM crops generate positive welfare gains to African countries south of the Sahara. The study also shows that welfare gains associated with adoption of GM crops outweigh the gains tied to greater market access in restrictive export destinations, such as the EU. The estimated gains would be slightly lower if the EU's policies continue to effectively restrict imports of affected crop products from countries that decide to adopt GMOs in Africa (Anderson and Jackson 2005).

These results are generally consistent with one of the other papers in the literature, and they do not seem to depend too much on the crop studied or the adoption level. Because of the low trade flows and the fact that the gains of GM crops are mostly captured domestically, African countries are found to gain regardless of regulatory barriers.

Yet at the same time, a few of these papers focus on potential trade implications within Africa. Anderson and Jackson (2005) is the exception. It shows that a GM ban in countries of the Southern African Development Community to save rents from exports to Europe results in lower gains than would be obtained by GM crop adoption.[3] Still, regional restrictions can matter, as shown by recent developments in South Africa. Some countries in southern Africa have adopted (transitional or seemingly established) strict import policies on GM food. This did not prevent South Africa from successfully adopting and marketing GM maize in the past (Gruère and Sengupta 2010), but the situation changed in 2010 with an unexpectedly large harvest, leading to excessive marketable surplus. The combination of high GM maize adoption and the continued presence of GM import barriers in the region resulted in large domestic stocks that traders have had trouble selling, despite a significant maize shortage in parts of East Africa. As a result, the South African maize price dropped, while Kenya and other countries had to purchase high premium non-GM maize for imports.

This contrasting example underlines the role of regional integration, which is the subject of the following section.

---

3 Langyintuo and Lowenberg-DeBoer (2006) study the trade effect of differential adoption of a GM cowpea in various countries of West Africa, but they do not model regulatory barriers.

The Case for Regional Integration

The Biotechnology Revolution and Regional Integration

Regional integration is a process in which states enter into a regional agreement to enhance cooperation through regional institutions and rules. Its objectives can range from economic to political, although it has become a political economy initiative where commercial purposes are the means to achieve broader sociopolitical and security objectives. Efforts at regional integration have often focused on removing barriers to free trade in the region, increasing the free movement of people, labor, goods, and capital across national borders. This works as a spur to greater efficiency, productivity gain, and competitiveness, not just by lowering border barriers, but also by reducing other costs and risks of trade and investment.

In numerous forums, African leaders have underscored the importance of greater coordination and harmonization among the continent's many regional economic communities. For instance, the most important of these efforts are the Abuja Treaty establishing the African Economic Community and the more recent Constitutive Act of the African Union. Article 3 of the Act of the African Union underscores the need to "coordinate and harmonize the policies between the existing and future Regional Economic Communities for the gradual attainment of the objectives of the Union" (AUC 2011). The New Partnership for Africa's Development, which is one of the core programs of the African Union, assigns a significant role to the regional economic communities, emphasizes regional and subregional approaches, and encourages African countries to pool resources to enhance growth prospects and to build and maintain international competitiveness (UNECA 2006).

According to the COMESA secretariat, creation of a Customs Union has increased trade from US$9 billion in 2007 to US$15.2 billion in 2008. To sustain the positive trend, regulating trade in products that contain or may contain GMOs and transboundary movements of GMOs across porous borders is going to be a fundamental element. It is on these grounds that African leaders have demonstrated political will and commitment to cooperate and take a common approach to biotechnology and biosafety issues at the regional level.

High-level meetings have been held to discuss matters touching on harmonization of biosafety standards. Deliberations at the Extraordinary Conference of the African Ministers Council on Science and Technology, held in Cairo in November 2006, focused on the African Strategy on Biosafety. The Strategy targets the national and subregional levels for planned interventions to be undertaken by the African Union and its member states to ensure harmony in policies concerning modern biotechnology and biosafety. The Strategy recognizes efforts that the Regional Economic Communities have exerted in the area of biosafety or interlinking trade and biosafety. These initiatives will be used to complement the efforts of the Union. The Conference of African Union Ministers of Agriculture, held in Libreville, Gabon, from November 27 to December 1, 2006, deliberated on *An African Position on GMOs in Agriculture* (AU 2006). The ministers recommended that the African Union set up mechanisms to identify commonalities among the Regional Economic Communities to be used to harmonize and coordinate policies on biosafety and biotechnology. Developments in regional harmonization are now conspicuous in the COMESA region, the Economic Community of West African States, and the Southern African Development Community.

Implications of Commercializing GMOs for Intraregional Trade in Africa

Currently, three African countries have granted approvals for commercial release of GMOs. South Africa has been growing insect-resistant (Bt) cotton, Bt maize, and GM soya for more than a decade. Burkina Faso and Egypt commercialized Bt cotton and Bt maize, respectively, in 2008.

The importance of such commodities as cotton and maize has been recognized in the region by such bodies as COMESA and the East African Community (EAC). Maize is a principal food-security commodity that dominates both formal and informal trade, especially in eastern and southern Africa. In 2007, the size of the regional maize market (EAC and COMESA) was estimated to be slightly more than US$1 billion. The total maize consumption in the region is about 16 million metric tons per year (Njukia 2006). The EAC development strategy for 2001–05 advocated liberalization of maize trade. COMESA, in collaboration with EAC and the Regional Agricultural Trade Expansion Support, has been exploring options for fostering regional maize trade. One of the strategies embraced is a "maize without borders" concept involving harmonized policies and regulatory frameworks to facilitate increased movement of maize across borders (COMESA 2003).

Efforts to promote trade in cotton in the COMESA and EAC regions have also been noted. In 2005, the first Regional Cotton and Textile Executive Summit was held in Nairobi, Kenya. The meeting resulted in the formation of the Africa Cotton & Textile Industries Federation (ACTIF) in the same year. It is a regional body composed of the cotton, textile, and apparel sectors from across Africa south of the Sahara aimed at creating a unified and recognized voice in both regional and global trade affairs. ACTIF was officially launched on March 28, 2010.[4]

As evidenced by profiles of various commodities in the recent past (ICTSD and ATPS 2007), exports of crops for which biotech varieties could potentially be introduced and commercialized in Africa are largely destined to other African countries. This is a strong indication that African countries should be more concerned about possible commercial export risks associated with intraregional trade. For instance, less than 5 percent of maize produced in Kenya, Tanzania, and Uganda is exported outside Africa. The share of maize traded intraregionally (that is, within Africa) is 98 percent for Kenya, 96 percent for Tanzania, and 99 percent for Uganda (ICTSD and ATPS 2007).

As borders are porous and effective mechanisms and infrastructure for monitoring transboundary movement of GMOs are often absent, Bt maize and Bt cotton can easily be shipped from one country to another either formally or informally. If Kenya commercialized Bt maize and Bt cotton ahead of Tanzania and Uganda, for instance, the chances of trade disputes erupting are high unless biosafety regulatory instruments are put in place to address transboundary movement of such commodities.

As mentioned earlier, Regional Economic Communities in Africa are pushing for increased regional integration and free-trade areas. However, with some countries moving ahead with the commercialization of GMOs and others considering adoption of a GM-free stance, the unrestricted and rapid movement of such commodities as maize may not be realized. At the national level, several countries have provisions for handling imports, exports, or GMOs in transit in their policies, laws, or regulations. However, it is important that such biosafety considerations are adequately incorporated into instruments of regional economic integration to safeguard disruption of intraregional trade. In the absence of mutually acceptable regional policies and guidelines, some countries may decide to go for stringent identification, testing, and labeling procedures. This would drastically slow down smooth cross-border flow of essential commodities. The costs of biosafety regulation could also increase the price of food in the importing countries.

Conclusions

In this chapter we discussed export risks associated with the use of GM crops in African countries. Several African countries have adopted a precautionary

---

4 See www.cottonafrica.com/news.php?newsid=5.

posture, to the point of imposing bans on adoption of GM crops, in an attempt to maintain access to target export markets (particularly countries in Europe) that prefer non-GM products. Yet our review of case studies, trade data, and simulation analyses suggests that most of the alleged risks are limited, especially when compared to the potential benefit of adopting GM crop technology.

Although each of these study approaches has clear limitations, their results indicate that the presence of an immediate risk of export loss has been exaggerated. In particular, an analysis of commodities produced and exported from African countries indicates that the magnitude of short-run commercial export risk in destinations outside Africa would be relatively small in monetary terms, because current GM crops (like maize) are largely traded intraregionally. Agricultural exports to the EU, Asia, and the Middle East are traditionally of cash crops or horticultural products, such as tea, coffee, vegetables, and flowers.

Still, because the results of these analyses are based on existing patterns of trade, and trade patterns may change (or have the potential to change with certain policies and investments), the consideration of the export consequences of allowing and expanding GM crops may differ across countries and would need to be assessed specifically for each country, depending on its circumstances and agricultural potential.

The example of South Africa is particularly telling; it was considered a successful importer and exporter of GM and non-GM maize in the past (Gruère and Sengupta 2010). But after an unexpectedly high GM maize harvest in 2010, it faced the problem of oversupply with insufficient foreign demand, driving domestic prices down, notably because of the import bans on GM products in the region.[5]

Thus, countries should carefully weigh the potential risks of losing export earnings on a case-by-case basis. Reluctance to approve beneficial GM crops, especially those that could deliver large welfare gains, is likely to deny farmers in the region the opportunity to harness the diverse benefits of GM crops and products. But rapid approval of highly traded GM commodities in the presence of foreign regulations and low excess demand may also prove detrimental, leading to price depression.

---

5 After government efforts to find external buyers, in 2011/12, South Africa faced the reverse problem of excessive external demand, forcing South African animal feed manufacturers to purchase maize from other countries at a higher price than the domestic price.

The importance of current intraregional trade in GM-sensitive commodities implies that trade disputes could also erupt—especially if some countries reject imports of agricultural commodities on the grounds that they could be contaminated with GM products. This would impact negatively on current efforts to facilitate unrestricted movement of commodities, such as the "maize without borders" concept and customs union arrangements that have been concluded or are being negotiated.

Thus, regional cooperation in matters related to the transboundary movement of GMOs is bound to be of great importance. Bilateral and multilateral mechanisms to manage intraregional trade in products that may contain GMOs is required to ensure that the goals of regional integration are not jeopardized. Biosafety considerations should be adequately incorporated in regional integration instruments.

References

Anderson, K., and L. A. Jackson. 2005. "Some Implications of GM Food Technology Policies for Sub-Saharan Africa." *Journal of African Economies* 14 (3): 385–410.

Anderson, K., and E. Valenzuela. 2007. "The World Trade Organisation's Doha Cotton Initiative: A Tale of Two Issues." *World Economy* 30 (8): 1281–1304.

AU (African Union). 2006. "An African Position on Genetically Modified Organisms in Agriculture." Presented at the Conference of African Union Ministers of Agriculture, November 27– December 1, in Libreville, Gabon.

AUC (African Union Commission). 2011. Third Publication African Union Commission, July. http://ea.au.int/en/sites/default/files/SIA_English.pdf.

Bouët, A., and G. Gruère. 2011. "Refining Estimates of the Opportunity Cost of Non-adoption of Bt Cotton: The Case of Seven Countries in Sub-Saharan Africa." *Applied Economics Perspectives and Policy* 33 (2): 260–279.

COMESA (Common Market for Eastern and Southern Africa). 2003. Thirteenth Meeting of the COMESA Trade and Customs Committee, October 15–17, in Lilongwe, Malawi.

Diao, X., M. Johnson, S. Gavian, and P. Hazell. 2005. *Africa without Borders: Building Blocks for Regional Growth.* IFPRI Issue Brief 38. Washington, DC: International Food Policy Research Institute. www.ifpri.org/sites/default/files/publications/ib38.pdf.

Elbehri, A., and S. MacDonald. 2004. "Estimating the Impact of Transgenic Bt Cotton on West and Central Africa: A General Equilibrium Approach." *World Development* 32 (12): 2049–2064.

Gruère, G. P. 2006. *An Analysis of Trade Related Regulations of GM Food and Their Effects on Developing Countries' Decision Making.* Environmental and Production Technology Division Discussion Paper 147. Washington, DC: International Food Policy Research Institute.

Gruère, G. P., and Sengupta, D. 2009. "GM-Free Private Standards and Their Effects on Biosafety Decision-making in Developing Countries." *Food Policy* 34 (5): 399–406.

———. 2010. "Reviewing South Africa's Marketing and Trade Related Policies for Genetically Modified Products." *Development Southern Africa* 27 (3): 333–352.

Gruère, G. P., and H. Takeshima. 2012. "Will They Stay or Will They Go? The Political Influence of GM-Averse Importing Companies on Biosafety Decision Makers in Africa." *American Journal of Agricultural Economics* 94 (3): 736–749.

ICAC (International Cotton Advisory Committee). 2010. *World Cotton Trade.* Washington, DC.

ICTSD and ATPS (International Centre for Trade and Sustainable Development and African Technology Policy Studies Network). 2007. *Biotechnology: Eastern African Perspectives on Sustainable Development and Trade Policy.* Geneva and Nairobi.

Juma, C. 2010. *The New Harvest: Agricultural Innovation in Africa.* Oxford: Oxford University Press.

Juma, C., and I. Serageldin. 2007. "Freedom to Innovate: Biotechnology in Africa's Development. A Report of the High-Level African Panel on Modern Biotechnology." Addis Ababa, Ethiopia, and Pretoria, South Africa: African Union and New Economic Partnership for Africa's Development. http://mrcglobal.org/files/Singh_Africa_20Year.pdf.

Langyintuo, A. S., and J. Lowenberg-DeBoer. 2006. "Potential Regional Trade Implications of Adopting Bt Cowpea in West and Central Africa." *AgBioForum* 9 (2): 111–120.

Njukia, S. 2006. "Using Markets to Increase Food Security." Presentation at the 2nd African Drought Risk and Development Forum, October 16, in Nairobi, Kenya. www.disasterriskreduction.net/fileadmin/user_upload/drought/docs/ADAF2006_1.8_Njukia.pdf.

Paarlberg, R. 2009. *Starved for Science: How Biotechnology Is Being Kept out of Africa.* Cambridge, MA, US: Harvard University Press.

Paarlberg, R., D. Wafula, I. Minde, and J. Wakhungu. 2006. *Commercial Export Risks from Approval of Genetically Modified (GM) Crops in the COMESA/ASARECA Region.* Nairobi, Kenya: African Centre for Technology Studies. www.asareca.org/paap/uploads/publications/Commercial%20Export%20Risks%20from%20Approval%20of%20GMO%20Crops.pdf.

Sinai, G. 2001. "Capital and Research Intensive Technology: New Monsanto and GMO Propaganda." *Le Monde Diplomatique*, July 2001.

Smale, M., P. Zambrano, G. Gruère, J. Falck-Zepeda, I. Matuschke, D. Horna, L. Nagarajan, et al. 2009. *Impacts of Transgenic Crops in Developing Countries during the First Decade: Approaches, Findings, and Future Directions.* IFPRI Food Policy Review 10. Washington, DC: International Food Policy Research Institute.

UNECA (United Nations Economic Commission for Africa). 2006. *Assessing Regional Integration in Africa. Policy Research Report.* Addis Ababa, Ethiopia.

Chapter 6

Estimates and Implications of the Costs of Compliance with Biosafety Regulations for African Agriculture

José Falck-Zepeda and Patricia Zambrano

As noted in the introduction to this volume, the area planted to genetically engineered (GE) crops has increased in developed and developing countries since their commercial approval in 1996 (James 2011). In Africa three countries, Burkina Faso, Egypt, and South Africa, have approved the commercial cultivation of GE crops. Regulatory systems in other African countries have approved the deployment of laboratory/greenhouse experiments and confined field trials for the development of more technologies. In addition, several developed and developing countries have made research and development (R&D) investments in crops and traits of local interest as a reaction to the observed benefits realized from GE crop cultivation in other countries (Atanassov et al. 2004). Yet public-sector research in developing countries, including in Africa, has released very few products to farmers, even though no damage to human health or the environment has been documented for any of the GE crops that have been approved and commercialized to date. This significant record is endorsed by science academies in Europe and the United States, international agencies, and national and regional regulatory agencies.

Two of the most pressing questions for Africa's national R&D systems and the International Agricultural Research Centers working in Africa are why so few GM products have made it to farmers and why there is such a limited number of crops, trait choices, and geographical locations for the release of potentially valuable technologies.

One answer may be that compliance with national biosafety regulations has increased the cost, time, and effort required for the approval of these technologies in African as well as in other countries. Increased costs may be negatively affecting research efforts to develop suitable and locally adapted technologies for developing countries (Kent 2004; Wright and Pardey 2006). Costs can be seriously increased when regulators are pressured to take into account concerns not necessarily related to the safety release of a product. For specific environments, valid safety questions might be raised regarding the effect of certain crops and traits. A solid regulatory system should be able to sort these valid concerns from those that are not needed to demonstrate safety.

As with any other innovation, conducting R&D to develop GE crops and releasing them to farmers is a long and costly investment project. Compared to conventional technologies, GE crops can carry higher R&D and technology transfer costs, but their stream of benefits over time can substantially offset these initial costs. Having a regulatory system that can operate in a timely and efficient way will guarantee that these benefits are not affected by unnecessary time delays, which is the single most important factor affecting benefit.

We structure this chapter as follows: first, we examine background biosafety issues related to the Cartagena Protocol on Biosafety; second, we discuss the estimation of the cost of compliance with biosafety regulations; third, we present estimates for the cost of compliance in selected countries, including those developed for the African context; fourth, we discuss the implications for national research systems in Africa and elsewhere; and finally, we provide some concluding comments.

Background on Biosafety Issues

The Cartagena Protocol on Biosafety does not explicitly define the term biosafety.[1] This situation is not unusual in international agreements, as there may not be sufficient consensus for an explicit definition in a multilateral context. The lack of definition leaves space for parties to the Protocol to define biosafety in their own legislation. Here we propose a functional definition of biosafety that considers the processes implemented within the scope of a regulatory system that enables a robust risk analysis of modern biotechnologies to ensure their safe use. Box 6.1 contains other biosafety definitions. Biosafety in this sense is a principle that evaluates the adoption of a new technology with careful consideration of its potential effects on the environment and human/ animal health. This definition is very broad and acknowledges the lack of a unique "best" approach to biosafety analysis (McLean et al. 2002). Each country bases its biosafety system on its own national, environmental, political, financial, and scientific capacities. The latter magnifies the importance

---

1 The Protocol does specify the scope in Article 4: "This Protocol shall apply to the transbound- ary movement, transit, handling and use of all living modified organisms that may have adverse effects on the conservation and sustainable use of biological diversity, taking also into account risks to human health" (SCBD 2000, 5).

BOX 6.1   Other Definitions of Biosafety

Convention on Biological Diversity—Biosafety Unit

Biosafety is a term used to describe efforts to reduce and eliminate the potential risks resulting from biotechnology and its products. For the purposes of the Biosafety Protocol, this is based on the precautionary approach, whereby the lack of full scientific certainty should not be used as an excuse to postpone action when there is a threat of serious or irreversible damage (see "What is the precautionary approach?"). While developed countries that are at the center of the global biotechnology industry have established domestic biosafety regimes, many developing countries are only now starting to establish their own national systems. [www.cbd.int/biosafety/faq/]

Ecolomics International

The concept of biosafety as it is used under the Cartagena Protocol on Biosafety specifies those legal actions that an importing country is entitled to take under international environmental law with the aim of protecting the biological diversity of its conventional plants and animals against the risk of contamination through imported varieties or species consisting of so-called Living Modified Organisms. These actions consist primarily of preventive or precautionary trade measures. Such restrictions or bans include the elaboration, negotiation and implementation of pertinent standards, and the institutionalization and international harmonization of the related regulatory framework and procedures. They also take into consideration the legally less clearly circumscribed concerns over related public health issues and socio-economic considerations. All these provisions aim at a non-hierarchical and mutually supportive relationship with other international agreements, especially with WTO [World Trade Organization] law, with the Codex Alimentarius, and with the International Plant Protection Convention. [www.ecolomics-international.org/headg_biosafety.htm]

---

of framing biosafety within each country's context as well as that of defining global principles of risk analysis and regulatory experiences.

In most countries with a biosafety system in place, regulators and policy-makers base their current biosafety assessments on past experience with risk analysis. These procedures provide a systematic and ideally science-based foundation that society can expand to address multiple consumer and other stakeholders' concerns and conflicting issues deemed important by society. Several countries, particularly those which signed and ratified the Cartagena Protocol, enacted regulations for assessing, managing, and communicating the risk assessment and subsequent decisionmaking for genetically modified crops (Mendoza 2005). A functional biosafety regulation system may have significant benefits for society and should ensure that the "right" technologies make it to market, while discarding those technologies that do not work or do not meet society's safety standard.

Most biosafety regulatory processes consist of a sequence of steps that require advance review and regulatory approvals by some combination of institutional, regional, or national biosafety committees. In Figure 6.1 we present an example of the typical steps taken for a biosafety assessment as part of a GE product development process. A typical sequence of steps may include laboratory trials, glasshouse/greenhouse (contained) trials, confined field trials, step-up (extended) field trials, and commercialization, although many variations of the sequence presented in Figure 6.1 exist.

The linear sequence of events in Figure 6.1 considers the possibility that each step builds on the accumulated knowledge in previous regulatory steps or on the generation of additional knowledge submitted to regulatory authorities. Biosafety regulators examine application dossiers submitted by the proponent considering the parent crop, the transformation method, the gene construct, and the GM crop for health and environmental impact. All activities conducted during the regulatory approval phase incur a cost. Therefore, proponents need to include biosafety regulatory costs in total development costs.

Risk and cost considerations bound biosafety assessments and biotechnology decisionmaking processes (Viscusi, Vernon, and Harrington 2000). The tensions among safety, technology use, and cost translate to trade-offs among

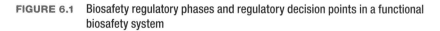

FIGURE 6.1   Biosafety regulatory phases and regulatory decision points in a functional biosafety system

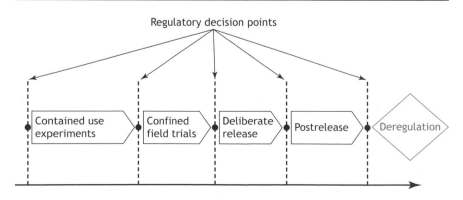

these competing issues. In terms of risk, society has a maximum level of risk that it is willing to tolerate while using innovations. In turn, the decisionmaking process is confined by cost considerations because of limited budgets for biosafety and biotechnology review processes. Policymakers may describe policy options through risk and cost combinations offering society the same level of risk at a lower cost or the same level of cost but with a lower level of risk. These trade-offs highlight the need for societies to define a decisionmaking path that will guide their actions.

The idea that risk, benefits, and cost trade-offs need to be examined during the design and implementation phases of biosafety and biotechnology regulatory processes is conceptually elegant and simple. However, actual risk assessment and eventual approval or rejection of a particular biotechnology is a complicated process. The safety profile of particular GE technologies consists of a portfolio of risk factors, each of which has a safety profile that may require its own decisionmaking process. For example, when examining food/feed safety, biosafety assessors may examine risk factors, such as allergenicity or toxicity, whereas when examining environmental safety, they may consider impact on non-target organisms.

The multifactor characteristic of the biosafety profiles of most GE technologies is further complicated as there may be different weights placed on each factor based on individual or institutional preferences. Regulators may deem the risk of one specific factor as unacceptable, whereas regulators may deem other risk factors as acceptable or manageable by risk-mitigation efforts. The multifactor characteristic of most GE technologies makes the cost–benefit and risk assessment much more difficult in practice.

Estimating the Cost of Compliance with Biosafety Regulations

Table 6.1 presents the four categories by which economists usually classify the social costs of regulations. These costs are listed in relative order of estimation from easiest to hardest. As an alternative, regulatory costs can be disaggregated into variable and fixed costs. Fixed costs are those that do not vary with the activity level of the research or production activity. In contrast, variable costs change with the level of activity. Regulations that impose high fixed costs on research institutions will hurt public-sector institutions (and small private firms), as they impose a higher per unit cost of research activity. Compared to private institutions, the public sector usually has more restrictive or less flexible budgets. Therefore, regulators need to consider "economies of scale" effects when designing and deploying biosafety regulations.

TABLE 6.1   Social costs of biosafety regulations

Social cost category	Examples
Real-resource compliance	Analytical tests Purchase (capital) new laboratory equipment Operation and maintenance of new laboratory equipment
Government-sector regulatory	National Biosafety Committees Training/administration Enforcement/litigation
Social welfare	Increased consumer and producer prices Additional legal/administrative costs
Transitional social	Unemployment Firms closing Public-sector institution abandoning research projects of public interest Transaction costs Disrupted production

Sources: Adapted from US EPA (2000) and Harrington, Morgenstern, and Nelson (1999).

Although it is worth considering all potential costs of implementing bio-safety regulations, some costs are relatively easy to conceptualize but very hard to quantify in practice. For example, the cost of rejecting projects that have potential social benefit will be difficult to measure. There are also other types of social costs that policymakers and regulators need to consider, such as transitional, social welfare, or indirect costs, which are also difficult to estimate. An example would be the opportunity costs of Institutional Biosafety Committees and National Biosafety Committees. What is important to take into account is the need to acknowledge that regulatory design and implementation has social costs and that there are trade-offs between innovation and regulatory intensity, especially if regulators move beyond what is needed to demonstrate safety.

Estimates of the Cost of Compliance with Biosafety Regulations

A handful of studies has documented the cost of compliance with biosafety regulations for a diverse set of countries and commodities. The cost evaluation used in many of the studies, particularly in developing countries with evolving regulatory frameworks, is mostly of an ex ante type. Hence, these studies derive the project costs of compliance from "best guess" estimates. For the ex post studies, the data collection approach considers those activities earmarked as biosafety regulatory costs by the developer. The biosafety regulatory costs in the studies presented here focus on the real-resource compliance costs and do not include other social costs, like government-sector regulatory costs, social welfare losses, and transitional and indirect costs.

The Insect-Resistant Maize for Africa Project in Kenya

A decade ago the International Maize and Wheat Improvement Center and the Kenyan Agriculture Research Institute, with the support of what today is known as Syngenta Foundation, launched the Insect-Resistant Maize for Africa (IRMA) project. The objective of the IRMA project was to develop through conventional and genetic-engineering methods maize varieties that are resistant to stem borers, a major pest in Kenya and other African countries south of the Sahara. In Kenya alone, estimates of this pest damage on maize production are approximately 14 percent, equivalent to 79 million US dollars (US$) (De Groote et al. 2004).

At the time of the project launch, biosafety regulations were only starting to take shape in Kenya. In 1998, a year before the launch of the IRMA project, the National Council for Science and Technology issued Regulations and Guidelines for Biosafety in Biotechnology and established the National Biosafety Committee (NBC), which is responsible for reviewing applications and setting guidelines for the Institutional Biosafety Committee.

In February 2000, scientists from the project submitted to the National Biosafety Committee their first application to import from Mexico leaf tissue from first-generation International Maize and Wheat Improvement Center events to screen their effectiveness against stem borers in Kenya. This required previous visits and inspections from the Kenyan Plant Health Inspectorate and the National Biosafety Committee (now known as the National Biosafety Authority of Kenya). Given that biosafety regulations were just starting to be put in place, it is not surprising that this application took more than a year to process (Mugo et al. 2005). After this learning experience, as illustrated in Table 6.2, the cost of these applications dropped substantially, and the processing time decreased over the years.

During the first 6 years, the project remained in the contained stage. The estimated costs of this stage were close to 1 million US dollars (US$), as detailed in Table 6.3. The IRMA project continues its activities to date with a renewed focus on conventional approaches to incorporating insect resistance to borers. The IRMA project has been a learning process for regulators and scientists in Kenya, who have become familiarized with a biosafety assessment process (Mugo et al. 2011). The project has opened the doors to other GE crops that national regulators have reviewed since the original biosafety

TABLE 6.2   Estimates of cost of applications over time (US dollars) and processing time (months) for the Insect-Resistant Maize for Africa (IRMA) project in Kenya

Cost category	Application for cut leaves, February (2000–2001)	Application for leaves, second generation (2002)	Application for seed (2003)	Application for field evaluation (2004–05)
Application preparation	59,200	19,767	15,226	22,840
Paper application	729	—	—	—
Meetings	6,944	—	—	—
Shipping materials	4,110	4,110	4,110	
Kenya Plant Health Inspectorate Service / regulator visits	—	833	1,111	1,750
Total cost	70,983	24,710	20,447	24,590
Processing time	13	9	5	3

Source: Authors' compilation for cost data and Mugo et al. (2005) for processing times.
Note: — = not applicable.

TABLE 6.3   Estimates of the cost of regulation for the containment stage of the Insect-Resistant Maize for Africa (IRMA) project in Kenya, 2005 (2005 US dollars)

Type of cost	Cost
Administrative (salaries)	7,000
Testing (overall)	416,000
Facilities and equipment	236,000
Capital expenditures	239,000
Overhead	68,000
Total	966,000

Source: Authors' estimates based on data from the IRMA project.

application for maize. In this sense, the IRMA example is quite atypical from what can be expected for other approval processes in Kenya and elsewhere.[2]

Bt Cotton in Kenya

In an effort to revive the cotton sector in Kenya, the introduction of transgenic varieties has been one of the policies that scientists and policymakers have

2 From the initial involvement of the National Council of Science and Technology in the biosafety regulatory process, which was somewhat ad hoc, Kenya now has a Biosafety Law signed into an Act in February 2009 by the president of Kenya. Furthermore, Kenya now has a functional National Biosafety Authority and thus a comprehensive and functional regulatory system with the involvement of multiple agencies.

TABLE 6.4   Estimates of the cost of regulations of Bt cotton in
Kenya, 2005 (2005 US dollars)

Stage	Initial estimated cost
Greenhouse trials	130,000
Controlled field trials	70,000
Scale-up	176,000
Commercialization	396,000
Total	772,000

Source: Authors' estimates.
Note: Bt = insect resistant.

been exploring for some years. Scientists estimate that the African bollworm
is the most damaging pest for cotton, and its control accounts for 30 percent
of production costs, with yield losses that can reach up to 100 percent (AATF
2006; Waturu 2006). The Kenyan Agriculture Research Institute has been
actively working on the introduction of insect-resistant (Bt) cotton varieties
since 2001, and the first confined field trials started in 2006. Table 6.4 presents the estimates for the cost of compliance with biosafety regulations for Bt
cotton in Kenya. Cost estimates in the case of Kenya add up to approximately
US$772,000.

Fungus-Resistant Bananas in Uganda

The National Research Organisation (NARO) of Uganda launched a joint
project with the Agricultural Biotechnology Support Program II and the
University of Leuven, Belgium, to develop a black Sigatoka–resistant banana
for the Ugandan context. The GE option is part of a broader effort launched by
NARO to develop bananas resistant not only to black Sigatoka but also to bacterial blights, nematodes, and other severe productivity constraints for Ugandan
agriculture by pursuing conventional and biotechnology means. The black
Sigatoka–resistant product entered a confined field trial in 2008, approved by
the National Biosafety Committee, under the supervision of NARO.

Table 6.5 includes an estimate of the potential costs of compliance with
biosafety regulations for the case of black Sigatoka–resistant bananas in
Uganda.[3] These costs are similar to those for a nematode-resistant vari-

---

3 A black Sigatoka–resistant banana developed by the abovementioned joint project is undergoing
   confined field trials in Uganda. It is not clear at this time whether this specific event will continue
   to the commercialization approval phase, or whether other events in development will be the can-
   didates for submission.

TABLE 6.5   Estimates of the cost of compliance with biosafety regulations for fungus- and nematode-resistant banana in Uganda, 2006–09 (2005 US dollars)

Stage/item	Description	2006	2007	2008	2009	Total cost
Laboratory	Application process	18,000	1,000	1,000	—	20,000
	Safety assessment	160,000	160,000	—	—	320,000
	Building regulatory capacity	17,200	6,000	6,000	—	29,200
	Development of biosafety facilities	130,400	58,500	58,500	—	247,400
Greenhouse	Cost of shipment and application	5,000	—	—	—	5,000
	Scientist's salary	24,000	—	—	—	24,000
	Labor	—	7,600	7,828	8,063	23,491
	Field materials and supplies	—	15,000	15,000	15,000	45,000
	Nursery and field evaluation of regenerated cell lines	—	5,000	5,000	5,000	15,000
	Containment facility	—	80,000	—	—	80,000
Confined field trial	Literature studies	—	—	10,000	—	10,000
	Fencing materials and construction	—	—	5,000	—	5,000
	Training technicians	—	—	10,000	—	10,000
	Security	—	—	1,200	1,200	2,400
	Salaries for two technicians	—	—	12,000	12,000	24,000
	Scientist's salary	—	—	24,000	24,000	48,000
	Other costs	—	—	5,000	5,000	10,000
Other	Equipment and facilities	10,139	10,139	10,139	10,139	40,557
	Overhead	33,630	41,190	15,304	10,539	100,664
Total		398,369	384,429	185,971	90,942	1,059,712

Source: Information from the National Agricultural Research Organisation, updated to 2009.
Note: — = not applicable.

ety in Uganda. Estimated costs through the confined field trials add up to approximately US$1 million. These estimates do not include the commercialization approval or the scale-up necessary to transfer the technology to farmers.

Costs for the black Sigatoka event are expected to be somewhat high compared to other events, due in part to the uniqueness of the event. There are very few instances of regulatory reviews of bananas and of fungal (or bacterial blight / nematode-resistance) traits. Nevertheless, if we compare the estimated costs stemming from regulations and R&D to the potential gains from the technology (see Chapter 4), we can affirm that it is prudent to assess properly this technology in order to enable the transfer of a safe and much-needed technology to farmers.

Estimates from Other Countries

Table 6.6 summarizes earlier estimates of regulatory compliance costs from different sources. Regulatory compliance costs varied across commodities and countries, ranging from US$500,000 for Bt cotton in India to US$4 million for soybeans in Brazil. These initial estimates of the costs of compliance with biosafety regulations were made based on the existing regulatory process at the time of data collection. In some cases, these estimates are likely to change with changes in the regulatory process; therefore, they need to be updated over time. Regulatory cost estimates in Table 6.6 for technologies in developing countries tend to be lower than the costs incurred by private companies for earlier technologies in the United States, as seen in Table 6.7.

As summarized in Table 6.7, estimates of regulatory costs in the United States in the private sector vary from US$7 million to US$15 million per single new product for first-generation products released into the country (Kalaitzandonakes, Alston, and Bradford 2007). There are several potential

TABLE 6.6   Cost of compliance with biosafety regulations

Type of crop	Crop	Country	Event approved in developed countries	Estimated cost of biosafety regulations (US$)
Foodcrop	Maize	India	Yes	500,000–1,500,000
	Maize	Kenya	Yes	980,000
	Rice	India	No	1,500,000–2,000,000
	Rice	Costa Rica	No	2,800,000
	Beans	Brazil	No	700,000
	Mustard	India	No[a]	4,000,000
	Soybeans	Brazil	Yes	4,000,000
	Potatoes	South Africa	Yes	980,000
	Potatoes	Brazil	—	980,000
	Papaya	Brazil	Yes	—
Non-foodcrop	Cotton	India	Yes	500,000–1,000,000
	Jute	India	No	1,000,000–1,500,000

Sources: Compilation from Falck-Zepeda et al. (2007), based on estimates from Odhiambo (2002), Sampaio (2002), Sittenfeld (2002), and Quemada (2003). Data for India from Pray, Bengali, and Ramaswami (2005).
Note: — = not applicable; US$ = US dollars.
[a]Producer must seek approval in export markets.

TABLE 6.7   Estimated costs for biosafety activities for India, China, and the United States (US dollars)

Activity	India	China	United States Minimum	United States Maximum
Pre-approval				
Molecular characterization	—	—	300,000	1,200,000
Toxicology (90-day rat trial)	—	14,500	250,000	300,000
Allergenicity (Brown Norwegian rat)	150,000	—	—	—
Animal performance and safety studies	—	—	300,000	840,000
Poultry-feeding study	5,000	—	—	—
Goat-feeding study (90 days)	55,000	—	—	—
Cow-feeding study	10,000	—	—	—
Water-buffalo feeding study	10,000	—	—	—
Fish-feeding study	5,000	—	—	—
Anti-nutrient	—	1,200	—	—
Gene flow	40,000	11,200	—	—
Baseline and follow-up resistance studies (ELISA)	20,000	—	—	—
Protein production/characterization	—	—	160,000	1,700,000
Protein safety assessment	—	—	190,000	850,000
Non-target organism studies	—	11,600	100,000	600,000
ELISA development, validation, and expression	—	—	400,000	600,000
Composition assessment	—	—	750,000	1,500,000
Limited, multilocational, and/or large field trials	585,000	—	130,000	460,000
Socioeconomic studies	15,000	—	—	—
Facilities, management, salaries, and other overhead costs	900,000		600,000	4,500,000
Total pre-approval	1,795,000	—	3,180,000	12,550,000
Post-approval				
Socioeconomic study	30,000	—	—	—
Resistance study	20,000	—	—	—
Integrated pest-management package	20,000	—	—	—
Facilities, management, salaries, and other overhead	125,000	—	—	—
Total post-approval	195,000	—	—	—
Total	1,990,000	—	3,180,000	12,550,000

Sources: Estimates for the United States from Kalaitzandonakes, Alston, and Bradford (2007); those for India from Pray, Bengali, and Ramaswami (2005); those for China from Pray et al. (2006).

Note: The data shown for the United States represent the average of the minimum and maximum for each activity across all responding firms in the author's survey. — = not applicable; ELISA = enzyme-linked immunosorbent assay.

explanations for the difference between the estimates of the cost of compliance presented here. These include cost underreporting by the public sector, lack of experience with earlier events, and the use of information generated in developed countries for use in developing countries. Some of these estimates are subjective and may be a reflection of uncertainties about the regulatory process that may be followed to ensure compliance.

Table 6.8 introduces estimates for the cost of compliance with biosafety regulations in the Philippines and Indonesia. Values presented in this table are for a combination of costs incurred to the time data were being collected in addition to the best estimate of potential costs to completion. Similar to the situation in Table 6.7, the estimates presented in Table 6.8 tend to be low compared to other estimates in the literature, in part because these are technologies approved in other countries, and thus developers were able to use—or are expecting to use—data generated elsewhere. Allowing the use of data generated elsewhere helps reduce compliance costs.

The case of Bt maize in the Philippines is different, in part because this technology was one of the first to go through the regulatory approval process in the country. As a result, there was very little regulatory experience with GM crops. Costs in the case of Bt maize are higher, because estimates in Table 6.8 consider the opportunity cost of using the information generated elsewhere by costing the relative value of maize to the

TABLE 6.8   Estimates of cost of compliance with biosafety regulations for selected technologies in the Philippines and Indonesia

Genetically modified crop	Country	Developer	Present value of total cost compliance with biosafety regulations (2005 US$)
Bt rice	Indonesia	LIPI	64,730
Drought-tolerant sugarcane	Indonesia	PTPN XI	94,389
Bt cotton	Indonesia	Monsanto	99,870
Herbicide-tolerant maize (RR NK603)	Indonesia	Monsanto	112,480
Bacterial blight–resistant rice (Xa21)	Philippines	PhilRice	99,213
Golden Rice	Philippines	International Rice Research Institute	104,698
Bt maize	Philippines	Monsanto	1,700,000
Delayed-ripening papaya	Philippines	Institute of Plant Breeding, University of the Philippines Los Baños	180,384

Sources: Falck-Zepeda et al. (2007, 2012).

Note: Bt = insect-resistant; LIPI = Indonesian Institute of Sciences; PTPN XI = Government Enterprise for Estate Crops (Indonesia); US$ = US dollars.

Philippines vis-à-vis global values (Manalo and Ramon 2007). If we disregard the opportunity costs for knowledge generated elsewhere, then we observe a reduced value comparable to the other crops shown in Table 6.8.

Values presented in Table 6.8 are similar to other estimates by Bayer, Norton, and Falck-Zepeda (2008). Their estimates for the Philippines show that the costs of compliance with biosafety regulations were similar or even larger than the development costs.[4]

Implications for Agriculture in African and Other Developing Countries

To date, only three countries have approved the commercial release of a GE crop (Burkina Faso, Egypt, and South Africa). Other countries have—or are—conducting confined field trials, including Kenya, Uganda, Tanzania, and Zimbabwe. Many more countries in Africa are considering developing their biosafety frameworks further to commercialize GE products after a biosafety assessment. In this section, we describe the potential impact of the cost of compliance with biosafety regulations for African agriculture. Because of the limited experience completing the different stages of a regulatory process that culminates with the potential commercial release of an assessed product, we rely on experience in other countries and sectors to discuss potential impacts.

High regulatory costs are especially relevant for regulatory impact assessment and technology decisionmaking efforts when they go beyond what is needed to demonstrate safety. Certainly, developers may conduct several activities during the regulatory stage that have a specific objective apart from biosafety, including identifying early marketable products or addressing potential liability concerns. Whether any cost is justifiable from a private or social investment standpoint will depend on its net contribution to social welfare. From the standpoint of regulatory impact assessment, concerns include inefficiencies, duplications, and unnecessary procedures that do not contribute to regulatory objectives. Furthermore, it is necessary to identify additional activities that yield enhanced safety knowledge but at an increasingly declining rate, as they may not justify more investments. Identifying these costs is paramount to improving the system.

---

4 Another study was done by Redenbaugh and McHughen (2004), who report that the cost of compliance with biosafety regulations in horticultural crops may be as low as US$1 million per allele but can be as high as US$5 million.

The high compliance costs observed so far relate to relatively well-described crops and traits, where science has accumulated a large body of knowledge on which to base regulatory assessments. Scientific and regulatory knowledge from these approved GE crops spilled over to other countries, thus reducing the potential cost needed to generate new information for similar or related technologies. Unfortunately, it is likely that current estimates for regulatory compliance costs represent a lower boundary for the expected costs for new crops and traits that will enter the regulatory pipeline in the near future. Although the opportunity arises of learning from existing regulatory processes to improve their efficiency, and the cost of performing individual scientific assessments may be declining over time, these trends may not be sufficient to balance the expected cost increase for novel products or traits entering the regulatory pipeline. Of course, the ability of knowledge to flow among countries provides a rationale for regional efforts to coordinate or harmonize assessments and decisionmaking processes.

Direct Cost Impacts

The high cost of compliance with biosafety regulations can affect public and private crop-improvement investments in different ways. The most obvious outcome is that a high compliance cost may reduce, ceteris paribus, investment in the development of regulated products and thus reduce the flow of potentially valuable products reaching farmers. In Africa, the impact of high regulatory costs may be more relevant to decisionmaking than elsewhere, as the likelihood exists that the regulatory systems in the continent may deal with crops and traits that have high social value in terms of benefits, but there may not be experience with or knowledge about the risk assessment.

These include crops like cassava, sweet potatoes, and bananas; or traits like viral and fungal resistance. However, risk assessment is a process that follows a set of well-defined steps identifying potential risk, likelihoods, and damage. This process has been validated through extensive risk-assessment experience not only with GE crops but also with pharmaceuticals, pesticides, and other chemicals. The scope then exists for developing robust systems in Africa that will respond to regulatory challenges in direct relation to the level of risk involved.

A more subtle outcome of high compliance costs is that, compared to unregulated products, public and private organizations may require a higher social/private rate of return for these regulated products as they carry greater R&D risks for developers. This outcome may have the unintended consequence of forcing developers to concentrate—and evaluate in more detail the potential impact—on higher return products, but because of budget constraints, African developers may forgo projects that might have high social returns over time but require a significant level of financial resources up front. Furthermore, high compliance costs may force the public sector with very limited annual budgets to refocus research efforts on unregulated products. In addition, the private (and even the public) sector may concentrate on those commodities with more innate private-return characteristics compared to crops of public interest. Certainly, the private sector focused on just four GE crops for the first generation of products after the marketability of their products was affected by regulatory costs and uncertainty.

Time Value of Money

A regulatory process extends the time needed to release a product for commercial purposes. Although this outcome is inevitable for regulatory processes, the core issue is to separate those regulatory activities needed to demonstrate safety from those added due to regulatory inefficiencies. Investments made during product conceptualization and R&D usually are done in advance of the regulatory process. Developers need to recuperate these investments as soon as possible because of the customary preference to have a return now rather than later. As time goes by, money itself "loses" its value at a discount rate determined by the time value of money to society or private interests.

Lack of clarity about regulatory requirements, scheduling, and other timing issues; restrictions on coordinating R&D and biosafety activities; limitations in terms of regulatory capacity of the competent authority; and other factors can compound the impact of regulatory delays. Delays beyond what would be needed to assess a product defer the onset of the stream of benefits generated through farmer adoption. Although there is very limited experience with assessing such delays in Africa, below we present a summary of studies in Uganda and in the Philippines, which show the critical need for reducing regulatory gaps and the importance of coordination to ensure prompt and efficient regulatory processes that deliver safe products.

The study by Kikulwe, Wesseler, and Falck-Zepeda (2008) examined the impact on social benefits from the introduction of a GE black Sigatoka–resistant banana in Uganda (see Chapter 4). This ex ante study incorporates irreversible cost and benefits using the real options approach. Results from this study show that if an approval delay occurs, Uganda forgoes potential social benefits of approximately US$179–365 million per year. Furthermore, assuming a maximum planting area of 541,530 hectares with a GE banana in Uganda, maximum total costs to bring the GE banana to Ugandan producers cannot exceed US$108 million. Otherwise, the GM banana is not a viable alternative. Results from this study show that regulatory delays and the total cost of development are substantial and need to be taken into account.

Figure 6.2 summarizes the results of the study by Bayer, Norton, and Falck-Zepeda (2008). This study examined the impact of increases in the cost of compliance with biosafety regulations and of regulatory delays on net benefits generated from the potential adoption of Bt eggplant, virus-resistant tomatoes, Bt rice, and virus-resistant papaya in the Philippines. Results showed that increases of up to 400 percent with respect to the

FIGURE 6.2   (a) Increase in the cost of compliance
                  (b) Compliance cost with increases in the time of approval

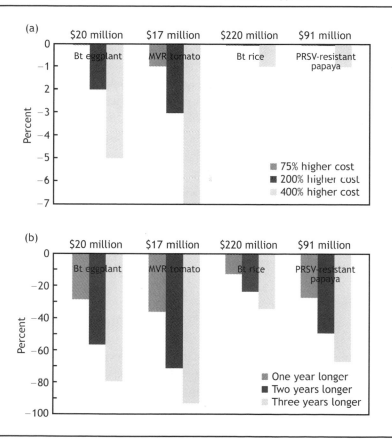

Source: Authors' estimates using Bayer, Norton, and Falck-Zepeda (2008).

Notes: Discount rate for the estimation of net present value is 5 percent. Change in net benefits is defined as the total benefits estimated using the economic surplus minus total regulatory costs. Bt = insect resistant; MVR = multiple virus resistant; PRSV = Papaya ringspot virus.

estimated base cost of compliance with biosafety regulations did not affect significantly the net present value of benefits from potential farmer adoption. In contrast, relatively small regulatory delays of 1–3 years reduced significantly the net present value of benefits from farmer adoption. This effect was especially significant in the case of tomatoes, eggplant, and papaya.

Entry Barriers and Regulatory Uncertainty

Parties to the Cartagena Protocol on Biosafety have chosen to pursue a biosafety assessment for GE crops before deliberate release into the environment. Most non-parties have also implemented a biosafety regulatory system based on existing regulations or through new national laws and regulations. Regulations carry financial and time costs that countries need to incur in order to guarantee the safety of their citizens. The problem arises when concerns not related to safety are added to these regulations, which turns the process into a serious hurdle for technology decisionmaking. In essence, the regulatory process may increase costs beyond what developers and society consider reasonable. In some cases, the turning point may be achieved when additional investments in biosafety assessments return reduced amounts of additional safety.

Excessively high regulatory compliance costs constitute a barrier to entry, especially for smaller organizations and the public sector when they are high enough to have an impact on investment decisions. This is true especially when organizations face budget constraints. Furthermore, in those countries where there is high regulatory uncertainty—in terms of not knowing how, if, and when a product may complete a regulatory process—the public and private sectors may reduce their levels of investment as they are less prepared to deal with investment uncertainty, which complicates decisionmaking under risk.

In a decisionmaking process under risk, R&D organizations typically contrast the cost of R&D and compliance with the probability of obtaining an approval for commercialization and the potential market returns from deploying a product; they then make an investment decision based on the risk-weighted returns. In the public sector, the focus is not on profit but rather on social return; thus, these organizations face a similar decisionmaking process.[5] In contrast, when R&D organization must make decisions under

---

5 Regulatory costs can also be considered as sunk costs once developers make investments in advance of knowing the outcome. Thus, developers would not factor them into the decisionmaking process once the investment is made. This old conundrum relates to whether analysts consider costs ex ante or ex post. There is no consensus among economists with regard to this issue.

uncertainty, the decisionmaking process becomes quite difficult and may represent a binding constraint for organizations examining investment for product development.

Although relatively high investments are needed to conduct biotechnology R&D, the costs appear to be declining significantly over time. In spite of the cost reductions for some activities, initial investments to develop biotechnology programs continue to be a barrier to entry for smaller organizations and the public sector in Africa and in other developing countries. The same holds for biosafety, as investment costs in buildings, equipment, laboratory capacity, and other capital investments can be onerous for developing countries. The pressing need exists to examine whether specific investments in biosafety are actually required to assess safety or the appropriateness of the technology development process, vis-à-vis the current innovation development stage in the country.

There are many strategies to reduce cost without sacrificing safety. Developers and regulators can use information generated in another country (or within the same country) in order to guide the biosafety assessment of a particular event. In addition, the regulatory system may reduce the required information needed by identifying activities that add no value for assessing safety. Assessors may consider identifying specific risk considerations that do not contribute to the risk profile of the technology. Alternatively, developers can conduct some of the more expensive food/feed safety tests in countries where it is less costly. As long as regulations or laws governing the process remain unchanged, costs will decrease as developers and regulators become more experienced and more events go through the regulatory process.

Sources of Funding

As mentioned earlier in this chapter, researchers should not stop their research projects on the basis of the high costs of compliance without taking into consideration the relative importance of these estimates to the total cost of development, the importance of the crop and its trait for national interests, and the estimated level of benefits that may be achieved by using the technology. The latter issue connects with the damage averted or gains made from using the technology. Furthermore, researchers need to take into account how compliance costs have changed over time.

In the absence of other considerations, if developers have a high-value product—especially those in the public sector with significant value for resource-poor farmers—then the relatively high R&D costs (including those for biosafety assessment) have to be contrasted with whether it pays to invest in such technologies and who will pay for it. In other words, if a particular technology demonstrates a real positive public value, then identifying the entity that will pay for bringing this technology to farmers is key in the decisionmaking process

Unfortunately, public research budgets face the current declining trends in R&D investments seen not only in Africa but also in other regions. This situation contrasts with the pressing needs that accelerating climate change and additional population pressure will bring to African countries, on top of the existing lags in innovation investment. Although the international research system and some developed countries have expressed their will to support ongoing innovation activities, support has not fully materialized.

Conclusions

Modern biotechnology needs to be a part of the tools used for effective poverty alleviation in Africa. As GE crops and other products are regulated products, it is imperative to establish regulatory systems that are commensurate with the potential risk of the technology. These systems should be not only flexible enough to adapt to gains in knowledge and experience, but also transparent and fair, and take into consideration all aspects of a broad and inclusive decisionmaking process. Biosafety systems that are too cumbersome or inflexible and those that become an insurmountable hurdle will stop this technology in its tracks, even those that have an elevated potential to resolve specific productivity issues of African agriculture. Biosafety thus becomes a process that considers all costs, benefits, and risks of prospective technologies within the scope of overall sustainable agricultural and economic development.

So far, GE crops that have been released deliberately have had a remarkable safety record. No proven damage to human health or the environment has been documented for any of the approved GE crops. Science academies in Europe and the United States, international agencies, and national and regional regulatory agencies officially endorse this safety record. However, novel crops and traits that will enter the regulatory pipeline may present a new set of challenges for regulatory agencies and biosafety systems.

Biosafety regulatory systems assess, manage, and communicate the objective risks posed by GE crops to human health, the environment, and biodiversity. Benefits and costs from the potential introduction of the technology have only received a cursory attention by most regulatory systems.

Furthermore, failing to adopt new technologies that may benefit poor farmers and consumers carries its own set of risks (Nuffield Council on Bioethics 2003). Undesirable conventional agricultural production practices may result in overexposure to chemical pesticides or incomplete abatement of pest damage, negatively affecting food safety and security. Biosafety regulatory systems thus need to balance objective risks to human health and the environment against the potential risk of opportunity losses to increase agricultural production, introduce novel crops, and enhance the livelihoods of poor people.

The key to delivering safe, valuable, and appropriate GE technologies to farmers in poor countries is smart and efficient regulatory systems implementable by countries with lower scientific and financial capacity. Developing countries need such a system, which does not necessarily mean more biosafety regulations. Regulations have to be based on risk-assessment procedures with a history of success in other countries. Many international documents articulate biosafety, risk assessment, and risk management principles; a specific example is Annex III to the Cartagena Protocol on Biosafety. Furthermore, establishing sensible regulations does not have to be complicated, as long as the process is robust, transparent, and participative (Jaffe 2006). Most importantly, biosafety needs to be a process trusted by society. Several policy and regulatory system options are available for improving biosafety processes.

African and other developing countries eventually will need to decide whether biotechnology and GE crops are useful for their development processes. This decision will be—hopefully—based on information on likely and actual impacts from biotechnology adoption and use (see, for example, Smale et al. 2009). Furthermore, developing countries need to realize that excessively precautionary approaches to biosafety regulations will have an impact on the flow of potentially valuable biotechnologies reaching farmers. All countries have the sovereign right to decide whether to embrace or reject this technology, but it is desirable that such decisions consider all costs, benefits, and risks involved with this technology. For those countries that are considering adopting GE crops, proper biosafety and technology assessments will help ensure the reduction of the costs and risks while maximizing the benefits.

Acknowledgments

The authors thank Hector Quemada, Karen Hokason, John Komen, and two anonymous reviewers for thoughtful and valuable contributions to this chapter.

References

AATF (African Agricultural Technology Foundation). 2006. Report of the Second Open Forum on Agricultural Biotechnology, October 26, in Nairobi, Kenya. Accessed November 2009. www.ofab.org/meeting_reports/OFAB2Report.pdf.

Atanassov, A., A. Bahieldin, J. Brink, M. Burachik, J. I. Cohen, V. Dhawan, R. V. Ebora, et al. 2004. *To Reach the Poor: Results from the ISNAR-IFPRI Next Harvest Study on Genetically Modified Crops, Public Research, and Policy Implications.* Environment and Production Technology Division Discussion Paper 116. Washington, DC: International Food Policy Research Institute.

Bayer, J. C., G. W. Norton, and J. B. Falck-Zepeda. 2008. "The Cost of Biotechnology Regulation in the Philippines." Selected Paper 469973 presented at the American Agricultural Economics Association Annual Meeting, July 27–29, in Orlando, Florida, US. http://purl.umn.edu/6507.

De Groote, H., S. Mugo, D. Bergvinson, and B. Odihambo. 2004. "Debunking the Myths of GM Crops for Africa: The Case of Bt Maize in Kenya." Paper presented at the American Agricultural Economics Association Annual Meeting, August 4, in Denver, Colorado, US.

Falck-Zepeda, J., J. Yorobe Jr., A. Manalo. G. Ramon, A. Bahagiawati, E. M. Lokollo, and P. Zambrano. 2007. "The Cost of Compliance with Biosafety Regulations in Indonesia and the Philippines." Selected Paper 175075 presented at the American Agricultural Economics Association Annual Meeting, July 29–August 1, in Portland, Oregon, USA. http://purl.umn.edu/9947.

Falck-Zepeda, J., J. Yorobe Jr., B. Amir Husin, A. Manalo, E. Lokollo, G. Ramon, P. Zambrano, and Sutrisno. 2012. "Estimates and Implications of the Costs of Compliance with Biosafety Regulations in Developing Countries: The Case of the Philippines and Indonesia." *GM Crops and Food: Biotechnology and Agriculture in the Food Chain* 3 (1). www.landesbioscience.com /journals/gmcrops/article/18727/?nocache=668315680.

Harrington, W., R. Morgenstern, and P. Nelson. 1999. "On the Accuracy of Regulatory Costs Estimates." Discussion Paper 99-18. Washington, DC: Resources for the Future.

Jaffe, G. 2006. "Regulatory Slowdown on GM Crop Decisions." *Nature Biotechnology* 24: 748–749.

James, C. 2011. *Global Status of Commercialized Transgenic Crops: 2011.* ISAAA Brief 43. Ithaca, NY, US: International Service for the Acquisition of Agri-Biotech Applications.

Kalaitzandonakes, N., J. M. Alston, and K. J. Bradford. 2007. "Compliance Costs for Regulatory Approval of New Biotech Crops." *Nature Biotechnology* 25 (5): 509–511.

Kent, L. 2004. "What's the Holdup? Addressing Constraints to the Use of Plant Biotechnology in Developing Countries." *AgBioForum* 7 (1–2): 12.

Kikulwe, E., J. Wesseler, and J. Falck-Zepeda. 2008. *Introducing a Genetically Modified Banana in Uganda: Social Benefits, Costs, and Consumer Perceptions.* IFPRI Discussion Paper 767. Washington, DC: International Food Policy Research Institute. www.ifpri.org/pubs/dp /ifpridp00767.asp.

Manalo, A. J., and G. P. Ramon. 2007. "The Cost of Product Development of Bt Corn Event MON810 in the Philippines." *AgBioForum* 10 (1): 19–32. www.agbioforum.org.

McLean, M. A., R. J. Frederick, P. L. Traynor, J. I. Cohen, and J. Komen. 2002. "A Conceptual Framework for Implementing Biosafety: Linking Policy, Capacity, and Regulation." ISNAR Briefing Paper 47. International Service for National Agricultural Research, the Hague, the Netherlands.

Mendoza, E.M.T. 2005. *Control of Ripening in Papaya and Mango by Genetic Engineering; Progress Report: May 2005.* Los Baños, the Philippines: Institute of Plant Breeding, College of Agriculture, University of the Philippines, Los Baños.

Mugo, S., H. de Groote, D. Bergvinson, M. Mulaa, J. Songa, and S. Gichuki. 2005. "Developing Bt Maize for Resource-Poor Farmers—Recent Advances in the IRMA project." *African Journal of Biotechnology* 4: 1490–1504.

Mugo, S., S. Gichuki, M. Mwimali, C. Taracha, and H. Macharia. 2011. "Experiences with the Biosafety Regulatory System in Kenya during the Introduction, Testing and Development of Bt Maize." *African Journal of Biotechnology* 10: 4682–4693.

Nuffield Council on Bioethics. 2003. *The Use of Genetically Modified Crops in Developing Countries.* London: Nuffield Council for Bioethics. www.nuffieldbioethics.org/go/ourwork/gmcrops/ publication_313.html.

Odhiambo, B. 2002. "Products of Modern Biotechnology Arising from Public Research Collaboration in Kenya." Presented at the International Service for National Agricultural Research conference Next Harvest—Advancing Biotechnology's Public Good: Technology Assessment, Regulation and Dissemination, October 7–9, in the Hague, the Netherlands.

Pray, C. E., P. Bengali, and B. Ramaswami. 2005. "The Cost of Regulation: The India Experience." *Quarterly Journal of International Agriculture* 44 (3): 267–289.

Pray, C. E., B. Ramaswami, J. Huang, R. Hu, P. Bengali, and H. Zhang. 2006. "Costs and Enforcement of Biosafety Regulations in India and China." *International Journal of Technology and Globalization* 2 (1–2): 137–157.

Quemada, H. 2003. "Developing a Regulatory Package for Insect-Tolerant Potatoes for African Farmers: Projected Data Requirements for Regulatory Approval in South Africa." Paper presented at the symposium Strengthening Biosafety Capacity for Development, June 9–11, in Dikhololo, South Africa.

Redenbaugh, K., and A. McHughen. 2004. "Regulatory Challenges Reduce Opportunities for Horticultural Crops." *California Agriculture* 58: 106–115.

Sampaio, M. J. 2002. "Ag-Biotechnology GMO Regulations/IP Progresses and Constraints." Presented at the International Service for National Agricultural Research conference Next Harvest—Advancing Biotechnology's Public Good: Technology Assessment, Regulation and Dissemination, October 7–9, in the Hague, the Netherlands.

SCBD (Secretariat of the Convention on Biological Diversity). 2000. *Cartagena Protocol on Biosafety to the Convention on Biological Diversity: Text and Annexes*. Montreal.

Sittenfeld, A. 2002. "Agricultural Biotechnology in Costa Rica: Status of Transgenic Crops." Presented at the International Service for National Agricultural Research conference Next Harvest—Advancing Biotechnology's Public Good: Technology Assessment, Regulation and Dissemination, October 7–9, in the Hague, the Netherlands.

Smale, M., P. Zambrano, G. Gruère, J. Falck-Zepeda, I. Matuschke, D. Horna, L. Nagarajan, et al. 2009. *Measuring the Economic Impacts of Transgenic Crops in Developing Agriculture during the First Decade: Approaches, Findings, and Future Directions*. Food Policy Review 10. Washington, DC: International Food Policy Research Institute.

US EPA (US Environmental Protection Agency). 2000. *Guidelines for Preparing Economic Analyses*. Washington, DC.

Viscusi, W. K., J. M. Vernon, and J. E. Harrington Jr. 2000. *Economics of Regulations and Antitrust*, 3rd ed. Cambridge, MA, US: MIT Press.

Waturu, C. 2006. "The Status of the Bt Cotton Confined Field Trials in Kenya." Accessed 2010. http://aatf-africa.org/userfiles/BtcottonKenya.pdf.

Wright, B. D., and P. G. Pardey. 2006. "Changing Intellectual Property Regimes: Implications for Developing Country Agriculture." *International Journal of Technology and Globalisation* 2 (1–2): 93–112.

Chapter 7

Policy, Investment, and Partnerships for Agricultural Biotechnology Research in Africa: Emerging Evidence

David J. Spielman and Patricia Zambrano

The past decade has seen vociferous debate over genetically engineered (GE) crops and their potential contribution to Africa's development. Although the initial debates revolved primarily around whether such technologies should be introduced, recent debates have become far more nuanced. Today, the discourse addresses such questions as what types of public policies can make GE crops and other applications of agricultural biotechnology (agbiotech) more readily available to researchers, farmers, and consumers in Africa. This chapter examines the extent to which public policies across the African continent are enabling the research, development, and dissemination of GE crops. The chapter does so by drawing on data from two studies on issues relating to the design and execution of such policies.

The first section reviews the literature on key issues in agbiotech research, development, and deployment, with specific reference to Africa. The next one describes the two studies on agbiotech in developing countries, with findings from these studies presented in the subsequent section. A set of actionable policy recommendations is then provided, followed by concluding remarks that offer insights to the future of GE crops and agbiotech in, and for, Africa.

Tracking Agbiotech's Evolution

The growing body of literature on the state of agbiotech in Africa includes three main areas of inquiry: (1) studies that track the development and commercialization of GE crops in the region; (2) studies that follow the development of biosafety regulations and related legislation for managing the release of GE crops in the region; and (3) studies of the size, types, and constraints to investment in research, development, and dissemination of GE crops by the public and private sectors. The literature in each of these areas is reviewed here.

Tracking the Introduction of GE Crops

The literature that tracks the development and application of GE crops is highlighted by several key sources. The first and most widely used source of information is the International Service for the Acquisition of Agri-biotech Applications (ISAAA), a global initiative that publishes periodic data and analyses of GE crop adoption worldwide. The information assembled by this initiative, along with its in-depth analyses of key trends and developments at the global, regional, and country levels (see, for example, ISAAA 2009), is possibly the most widely referenced source on these particular topics. A recent ISAAA global overview by James (2011) shows that in 2011, 160 million hectares of land across 29 countries worldwide were under GE crop cultivation, and the number of developing countries that had adopted GE crops totaled 18 (Argentina, Brazil, Bolivia, Burkina Faso, Chile, China, Colombia, Costa Rica, Egypt, Honduras, India, Mexico, Myanmar, Pakistan, Paraguay, the Philippines, South Africa, and Uruguay). Of these 18 countries, only 3 were in Africa: Burkina Faso, Egypt, and South Africa. In fact, South Africa was the only African country listed by ISAAA as planting GE crops up until 2008, when Egypt and Burkina Faso joined the list with the commercialization of Bt maize and Bt cotton, respectively.

A second source for GE crops is the Food and Agricultural Organization of the United Nations online Database of Biotechnologies in Use in Developing Countries (FAO-BioDeC). This database monitors trends in the development, adoption, and application of agbiotechs in developing countries (FAO 2011). For crop biotechnology FAO-BioDeC currently lists 3,104 products, of which 1,092 are classified as genetically modified organisms (GMOs) in a total of 50 countries. For countries in Africa south of the Sahara, only 11 are listed, with a total of 66 GMO crop products (Table 7.1). Most products are listed under South Africa, Kenya, and Nigeria.

A third source is the Center for Environmental Risk Assessment (CERA) database, formerly hosted by Agbios, a Canadian initiative that assembled detailed information on the global state of GE crop approvals by specific event (Agbios 2009).[1] The latest CERA database provides a list of 144 events that have been approved by national regulatory bodies from 23 countries and the European Union. In line with the data from James (2011), the only countries listed in CERA from Africa south of the Sahara are South Africa and Burkina Faso — South Africa with 25 approved products for cotton, maize, canola

---

1 "Event" refers here to the stable incorporation of foreign DNA into a living plant cell to create a potentially unique crop and trait combination.

TABLE 7.1   Number of genetically modified products in Africa south of the Sahara, 2011

Country	Number of products
South Africa	40
Kenya	9
NIgerIa	5
Uganda	3
Zimbabwe	2
Burkina Faso	2
Tanzania	1
Cameroon	1
Ghana	1
Malawi	1
Mauritius	1
Total	66

Source: Adapted from the online database Biotechnology in Developing Countries of the Food and Agriculture Organization of the United Nations (FAO 2011).

and soybean, and 1 for Burkina Faso with the recently commercialized cotton (CERA 2011).

A fourth source is bEcon, a web-based bibliography maintained by the International Food Policy Research Institute (IFPRI) of the applied economics literature that documents the impacts of GE crops in developing countries (Yerramareddy and Zambrano 2011). Whereas other sources of information focus primarily on the commercialization of GE crops and the regulatory regimes that promote or impede commercialization, bEcon and an accompanying literature by Smale et al. (2009) offer additional insights into the costs and benefits of agbiotech to farmers and consumers.

Analysis from these sources also suggests a mixed record of impact to date. Although James (2011) and others are generally optimistic about the gains from the rapid diffusion of insect-resistant (Bt) cotton and other trait/crop combinations, Smale et al. (2009) offer a mixed report on economic outcomes. They caution that positive findings in terms of cost reductions and yield improvements for Bt cotton in China and India should not be generalized to other traits, crops, and countries, and that more extensive and comprehensive research is needed to better understand the impacts of GE crops in developing countries.

Tracking Biosafety Legislation and Regulation

Studies of developments in the field of biosafety legislation and regulation in Africa are generally in-depth examinations on governance issues relating to the research, development, and dissemination of GE crops to researchers, farmers, and consumers in Africa. Key studies in this area include Baum, de Kathen, and Ryan (2001); Johanson and Ives (2001); Mugabe (2002); Alhassan (2003); Sithole-Niang, Cohen, and Zambrano (2004); Harsh (2005); Wafula and Clark (2005); and Karembu, Nguthi, and Abdel-Hamid (2009). These studies assess a range of regulatory issues, such as the health and environmental risks of GE crop cultivation, the costs and benefits of agbiotech regulation, transparency and accountability issues in regulating GE crops, and challenges related to capacity strengthening to support the design and implementation of biosafety regulations.

Studies on the broader governance issues, including the political and political economy aspects of agbiotech and their influence on national regulatory systems in Africa (and in other developing countries) are offered by Komen, Webber, and Mignouna (2000); Paarlberg (2001); Cohen and Paarlberg (2004); Cohen (2005); Pray, Bengali, and Ramaswami (2005); and Birner and Linacre (2008), among others. A conclusion that can be drawn from these studies is that timely advancement of agbiotech and GE crop research, and its effective use to address local problems in agriculture, is often hampered in countries where approvals are few and far between, where political interest groups advocate against the design and implementation of agbiotech policies and regulations, or where the capacity of the country's research and regulatory systems is limited.

Interestingly, although data reported earlier indicate that research and approval of GE crops in Africa is progressing at a sluggish rate, there is evidence suggesting that more progress has been made regarding the establishment of functional regulatory frameworks. In 2004, only five countries (Egypt, Kenya, Nigeria, South Africa, and Zimbabwe) had national biosafety policies in place: by 2009, this figure had increased to include Burkina Faso, Cameroon, Kenya, Malawi, Mali, Mauritius, Namibia, South Africa, Tanzania, Togo, and Zimbabwe (Karembu, Nguthi, and Abdel-Hamid 2009). Investment in the development of viable regulatory systems and the capacity to manage them has received a boost from the international community, with significant flows of resources originating from the US Agency for International Development (though the Program on Biosafety Systems and Agricultural Biotechnology Support Project II), the United Nations Environment Program–Global Environmental Facility, the World Bank, and others.

Regional initiatives and organizations in Africa—the Forum for Agricultural Research in Africa, the Common Market for Eastern and Southern Africa, the Association for Strengthening Agricultural Research in Eastern and Central Africa, the Economic Community of West African States, the West and Central African Council for Agricultural Research and Development, the Permanent Inter-State Committee for the Fight against Drought in the Sahel, and others—have also played a central role in strengthening capacity for biosafety regulation and facilitating the regional harmonization of biosafety regulations. These advances have been made in spite of the significant variation in how individual countries frame and execute their biosafety regulations (Spielman, Cohen, and Zambrano 2006a,b).

Tracking Public and Private Investment in Agbiotech

Studies of the levels, types, and constraints on investment in research, development, and dissemination of GE crops are a barometer of what is coming through the pipeline for Africa. Studies on public expenditure on agricultural research and development (R&D) are particularly useful in this context. Drawing on data from the Agricultural Science and Technology Indicators initiative online database, Pardey et al. (2006) find that most African countries faced declining or stagnating growth rates of public investment in agricultural R&D between the 1970s and 1990s. Beintema and Stads (2006), using data from the same database, estimate that the growth rate of public expenditure on agricultural R&D had declined from 2.0 percent in the 1970s to only 0.8 percent in the 1990s. Excluding South Africa and Nigeria from their sample—where R&D expenditures grew during the 1990s—total spending in the region actually declined by 0.2 percent per year, resulting in a halving of average spending per scientist (Beintema and Stads 2006). A complementary study by Falck-Zapeda et al. (2003) suggests that R&D specifically related to agbiotech represents a tiny fraction of these figures; it is concentrated primarily in South Africa, Kenya, and Egypt and is often highly dependent on donor funding. More recent evidence from Beintema and Stads (2011) indicates that although public R&D spending growth has recovered in the region between 2001 and 2008, growth has been concentrated in only a few countries and generally driven by staff salary increases and infrastructure rehabilitation, with little indication of growth in biotechnology investments.

Studies of private investment in the research, development, and dissemination of GE crops are also an important barometer of what is in store for

Africa. At the global level, investment in agbiotech research is not insignificant. Estimates suggest that private investment in plant biotechnology by the leading multinational companies during the mid-1990s totaled approximately 1 billion US dollars (US$) per year, a figure that amounts to roughly half of all global expenditure on agbiotech R&D (Byerlee and Fischer 2001; Pray and Naseem 2003). However, most of these expenditures were concentrated on crops, traits, and technologies directly relevant to industrialized country farming. Again, only a minute fraction of this expenditure is immediately relevant to Africa, a finding that Spielman (2007) suggests does not bode well for small-scale, resource-poor farmers in the region, although Pingali and Traxler (2002) are more sanguine.

But at a more local level, private investment in agbiotech is defined and constrained by the effectiveness and efficiency of seed systems and markets—beginning from varietal approval processes and ending with the distribution of improved cultivars to farmers. Functional seed systems are critical, because they are the channel through which many agbiotech applications will be deployed and disseminated in Africa. However, seed systems are, by their nature, subject to a variety of unique market and institutional constraints (Tripp and Louwaars 1997; Gisselquist and Van der Meer 2001; Tripp 2001).

First, problematic property rights result from the fact that improved seeds can, in many cases, be reproduced by the farmer, thus reducing the ability of breeders to appropriate the gains from their innovative activities and investments. Second, information asymmetries result from the inability of farmers to make ex ante assessments of seed quality, because such knowledge is held only by the seller in the absence of certain types of regulation. Third, coordination problems result from difficulties of enforcing and monitoring contracts for seed use: farmers often save and exchange seed without the breeder's knowledge. Finally, inelastic supply responses result from the inability of breeders to respond effectively to rapid changes in seed demand from farmers: often, farmers may reassess their seed type and quantity requirements just prior to planting based on expectations of rainfall, market prices, and other factors—decisions taken long after breeders have bulked up seed quantities for distribution.

Although seed systems may exist in Africa where markets for a given crop are well developed (for example, maize in Kenya), they are more often weak or otherwise incomplete due to the constraints noted above. Seed systems for "orphan" crops of marginal commercial value but of critical importance to subsistence farming—for example, sorghum, millet, groundnuts, pigeonpea, cassava, or sweet potatoes—are rarely functional in the region, primarily because of the combined weakness of the markets for these commodities, the nonappropriability of varietal improvements under current technological and legal regimes, and limited incentives to commercialize public research on varietal improvements in most countries (Tripp 2000, 2001). Thus, the market and institutional failures pose a significant barrier to the entry and growth of private seed firms that could potentially commercialize, market, and distribute varietal improvements resulting from public research. This is a major disincentive to increasing investment in GE crops and agbiotech research across the region.

One way of potentially bridging these systemic and market failures is to simultaneously draw on the assets and experiences of both the public and private sectors to develop and deploy GE crops in Africa. Studies by Pray (2001); Dubock (2003); Hall (2005); Spielman and Von Grebmer (2004); Chataway (2005); and Spielman, Hartwich, and von Grebmer (2007, 2010a,b) suggest that public–private partnerships (PPPs)—broadly described as any activity in which public and private entities jointly plan and execute activities with a view to accomplishing mutually agreed-on objectives while sharing the costs, risks, and benefits incurred in the process—represent an innovative approach to promoting agbiotech and GE crop R&D in developing countries.

R&D partnerships rely on processes of knowledge sharing, resource pooling, cost minimization, scale economies, and joint learning to generate synergies in conducting advanced research, commercializing new technologies, and deploying new products. Ideally, these synergies result in research outcomes of greater quantity; with a greater chance of success; or at lower cost than public, private, or civil society actors could expect when acting independently. If the research is strategically focused on the needs of marginalized social groups, outcomes may ultimately translate into significant social and economic benefits.

Partnerships are particularly useful to larger or more advanced systems that require access to cutting-edge research tools, proprietary knowledge, or other types of information and data; and to smaller systems that do not have the scale economies to conduct independent research efficiently (Byerlee and Fischer 2001). In recognition of this potential, key public-sector actors are engaged in several partnerships focusing on enhancing yields or nutritional content of such crops as rice, wheat, and cassava. In Africa, PPPs include research projects on Bt maize, water-efficient maize, Bt cowpeas, and disease-resistant bananas and plantains (see AATF 2009). PPPs in livestock vaccine development have also played a role in bringing agbiotech to bear on Africa's development; public–private research networks also exist for cassava and other crops (Smith 2005; Aerni 2006; Spielman 2009).

In many cases, members of the CGIAR Consortium play an important role in convening these projects and networks; mobilizing resources; and conducting research in close partnership with national research systems, multinational crop-science firms, and local seed firms. These projects and networks are expected to deliver beneficial outcomes over the next decade, although it remains unclear to what extent the partnership approach is yielding the anticipated outcomes.

Data, Data Sources, and Methods

This chapter examines agbiotech and GE crop research using data from two separate studies on agricultural R&D. The studies—each backed by its own survey instrument and data—provide some new evidence on the policy environments in agbiotech and GE crop research in Africa.

Next Harvest 2002

The first study, titled Next Harvest, was initiated in 2002 by IFPRI and the International Service for National Agricultural Research (see Atanassov et al. 2004). It was conducted to determine expectations and limitations on publicly researched GE crops and traits. The study was conducted based on an expert survey distributed to a purposeful sampling of 76 researchers and regulators working in public organizations in 16 developing countries. The sample was designed to capture the extensive variation in the type and state of research in different countries and organizations, and to ensure that relevant knowledge, experiences, and insight were provided to participants. Information on 209 GE crop products under development was received through the year 2003, along with the type of transgenes deployed, techniques used to deploy transgenes, types and sources of germplasm used, stage of regulatory approval reached, type of collaboration used to conduct the research, and plans for dissemination of research outputs. Fifty-four of the 209 products (26 percent) were attributable to GE research in African countries (Table 7.2).

Public–Private Partnerships 2006

The second study was undertaken by IFPRI in 2006 to examine partnerships between private firms and the members of the CGIAR Consortium

TABLE 7.2   Number of genetically modified crops under
development in Africa, 2003 and 2009

Country	2003	2009
Egypt	17	12
Kenya	4	5
Nigeria	n.a.	1
South Africa	28	15
Uganda	n.a.	4
Zimbabwe	5	2
Total	54	37

Sources: Atanassov et al. (2004) and Karembu, Nguthi, and Abdel-Hamid (2009).
Note: n.a. = Not available.

(see Spielman, Hartwich, and von Grebmer 2007, 2010a,b). The study
used four specific tools—document analysis of PPP-related materials, semi-
structured interviews with key informants engaged in partnership-based projects, an email survey of CGIAR centers, and development of a functional
typology of partnerships—to identify and analyze 75 CGIAR partnerships
that were active in 2004 (Table 7.3). A total of 12 out of 15 centers responded
to the survey and follow-up queries that were focused on the purpose, partners, outcomes, duration, and budgets of center PPPs. The survey also used
these tools to provide a more in-depth analysis of 6 partnership-based projects
at 4 separate CGIAR centers. A total of 14 projects (19 percent of all PPPs) in
5 centers (33 percent of all centers) involved agbiotech in some sense, with half
of these partnerships engaging leading crop-science companies in the sector.
Only 2 of these agbiotech projects were being conducted in Africa, although
the remaining 12 projects covered crops and traits that were also potentially
relevant to Africa. The first project is the Insect-Resistant Maize for Africa
project undertaken by the International Maize and Wheat Improvement
Center, Syngenta Foundation, Kenya Agricultural Research Institute, and
others. The second is the East Coast Fever vaccine development project
undertaken by the International Livestock Research Institute, Merial, Kenya
Agricultural Research Institute, and others.[2]

---

2 Since 2006, one new PPP involving agbiotech—the Water-Efficient Maize for Africa project—
has been launched. This brings the total number of PPPs in agbiotech research in Africa to three.
Other agbiotech PPPs may also exist in the region; however, no new survey has been conducted
since 2006 to document them.

TABLE 7.3   Distribution of public–private partnerships in the CGIAR, by center, as of 2004

International agricultural research center	Number	Share of total (%)
International Rice Research Institute[a]	17	23
International Crops Research Institute for the Semi-Arid Tropics[a]	11	15
International Center for Tropical Agriculture	10	13
International Maize and Wheat Improvement Center[a]	9	12
Bioversity International	8	11
International Center for Agricultural Research in the Dry Areas[a]	6	8
International Institute of Tropical Agriculture[b]	5	7
International Livestock Research Institute[a]	4	5
International Water Management Institute[c]	3	4
World Agroforestry Centre	3	4
International Potato Center	1	1
International Food Policy Research Center	1	1
Africa Rice Center	1	1
World Fish Center	0	0
Center for International Forestry Research[d]	0	0
Total	75	100
Total number of public–private partnerships in agricultural biotechnology	14	19

Source: Spielman, Hartwich, and von Grebmer (2007).

Note: A total of 75 partnerships were identified through the survey and other sources; four of these are multicenter partnerships. Because of multicenter partnerships, the entries for number and share total to more than 75 and 100 percent, respectively.

[a] A CGIAR center engaged in a partnership involving agricultural biotechnology.

[b] Did not provide survey responses. Information was obtained through document analysis.

[c] Did not provide survey responses. Information was obtained through document analysis and key informant interviews.

[d] Did not provide survey responses. Information could not be obtained by any method.

Key Findings

Findings from these two studies indicate that even though a growing number of countries in Africa are developing the necessary regulatory systems to support agribiotech research and GE crops, the actual research, development, and deployment of such products is lagging, largely because neither public- nor private-sector resources are being brought to bear in Africa, thus slowing the pace of innovation. These findings confirm those from other studies mentioned earlier, but they delve somewhat deeper into both the causes and consequences. This is examined in detail below.

Insights from Next Harvest

A key finding of the 2004 Next Harvest study was that public research institutions in developing countries have conducted a significant number of diverse crop transformations to express a wide variety of crop groups and transgenes.[3] However, although relatively large numbers of products were recorded in Asia and Latin America, the only African countries with any significant number of transformations were Egypt and South Africa. The situation has not changed substantially over the past 5 years, although a few countries have been added to the set of countries developing these technologies.

PRODUCT DEVELOPMENT

When classified by crop type, more than half (55 percent) of all public transformations recorded by the Next Harvest study were concentrated among 15 crops that are critical to achieving sustainable food security and reducing poverty in developing countries. The remaining 45 percent of products were focused on cotton, vegetables, and fruits—crops of a more commercial nature. For Africa, the predominant crop group in all 54 products was cereals, followed by vegetables, roots and tubers, and sugar, with each group representing a fairly diverse set of crop species. The greatest numbers of products among all 11 crops were for maize (17.0 percent), potatoes (13.0 percent), and sugar and tomatoes (11.0 percent each).

With regard to regulatory progress, most of these products remained confined to the experimental stage of laboratory and greenhouse trials; fewer have advanced to later stages in the regulatory process, such as field trials for biosafety testing, scaling-up for wider environmental and efficacy testing; or commercialization for release to farmers. Overall, African countries lagged slightly behind their Asian and Latin American counterparts: whereas 70 percent of all products in the African countries surveyed were still at the experimental stage, only 60 percent were at a similar stage in Asia and Latin America.

MAIN RESEARCH ACTORS

Most of the surveyed public organizations worked in isolation from other research actors, both public and private. In the study, only 7 percent of transformation products generated by these organizations were conducted in collaboration with the private sector, and only 22 percent were generated in

---

3 Findings based on data from the Next Harvest survey are reported in Atanassov et al. (2004); Sithole-Niang, Cohen, and Zambrano (2004); Cohen (2005); and Spielman, Cohen, and Zambrano (2006a,b). However, the findings presented here update these findings with additional analysis of the data and offer new insights from more recent research.

TABLE 7.4   Number of institutional arrangements used in public genetically modified products under development, by region and type of arrangement

Institutional arrangement	Africa	All regions
Single public institution	28	129
Public/public	13	47
Public/private	7	15
Public/foundation/public	0	8
Public/private/other	5	6
All other (no private collaboration)	1	4
Total	54	209

Source: Spielman, Cohen, and Zambrano (2006a).

collaborations between or among public institutions (Table 7.4). In Africa, the distribution of involvement was somewhat different: half (52 percent) of all transformation products were from a single public institution, whereas the others were from public–private (22.5 percent), public–public (13 percent), or some other type of collaboration (2 percent). Africa also had more representation from the private sector regarding origin of genetic materials. Although only 5 percent of all surveyed transformation products relied on genetic materials derived from local or foreign private-sector materials, 15 percent of all materials used in Africa originated from the private sector.

TRAITS OF GE CROPS

In terms of transgenes and gene groups, the figures suggest that agbiotech and GE crop research may be limited in focus with respect to the particular biotic and abiotic stresses facing agriculture in many developing countries. Fungal, bacterial, and other types of resistance are still at very preliminary stages of research for developing-country crops and agroecologies, whereas herbicide tolerance, insect resistance, and virus resistance—originally designed for the needs of industrialized-country agriculture—continue to dominate the research pipeline.

REGULATORY PROCESSES

In terms of regulatory progress, the figures indicate that forward movement in agbiotech and GE crop research in Africa is limited to very few countries and that research in those countries is only now reaching the initial stages of the regulatory process. Even though agbiotech may shorten the time needed to identify transgenes and transform plants, the resulting GE crop still requires time for scaling up, efficacy trials, environmental testing, and other regulatory requirements particular to genetic modification. Agbiotech research in Africa has not moved far along this road.

This reality is, according to many survey respondents, worsened by the fact that some countries have subjected GE crops to multiple years of testing—resulting in significant waiting periods for approvals for scale-up or pre-commercial trials—or have only interim guidelines or regulations in place that do not allow for commercial approvals. Even those countries that do have the ability to evaluate GE crops and provide commercial approvals often lack confidence in their commercial decisionmaking. Others may be facing such limitations as growers' inability to produce adequate amounts of seed for large-scale or food-safety testing.

PUBLIC- AND PRIVATE-SECTOR INTERACTION

The relatively small role attributable to the private sector in agbiotech and GE crop research in African countries suggests that public–private research collaborations face significant barriers to implementation. This absence of collaboration could pose difficulties for public institutions as they advance crops from research to regulatory approval and commercialization. Without exchanges of valuable regulatory data from private firms and other research institutions that have conducted transformations of similar crops and/or traits in industrialized countries, public institutions are poorly equipped to navigate the regulatory and commercialization processes with full information. Without scientific interaction and information exchanges between sectors, many of the public researchers who will be tapped for biosafety committees, regulatory agencies, or advisory bodies will be similarly less qualified to provide real expertise.

Insights from the IFPRI Study on PPPs

The 2006 IFPRI study on PPPs provides additional useful insights into interactions between the public and private sectors. This section highlights findings that relate to the main actors and their objectives, project costs and benefits, risk and risk-management strategies, and safe stewardship.[4]

MAIN ACTORS AND OBJECTIVES

A key finding of the 2006 IFPRI study was that multinational or foreign firms are engaged in only half of the 14 PPP-based projects that involved

---

4 The findings are based on data from the 2006 IFPRI study as reported in Spielman, Cohen, and Zambrano (2006a,b); Spielman, Hartwich, and von Grebmer (2007); and Spielman (2009). However, the findings presented here contain new analyses of the data and new findings that build on the previous work.

agbiotech, with the rest of these partnerships engaging local (typically seed) firms. And where multinational or foreign firms were engaged with a CGIAR center, the majority of such partnerships tended to be monogamous in nature, that is, involving only the center and the firm, without additional participation from other research organizations or firms. The projects involving multinational or foreign firms (for example, Monsanto, Syngenta, and Pioneer Hi-Bred International) were primarily designed to facilitate technology transfers and negotiate the use of intellectual property owned by the private-sector partner. In general, these projects did not leverage other private-sector assets, such as scientific expertise in working with agbiotech research tools or expertise in navigating regulatory processes to bring research into commercial use. In other words, few CGIAR centers engaged the private sector to conduct frontier research or to form ventures where public and private actors jointly undertake cutting-edge research activities characterized by some unknown probability of success. In Africa, the International Maize and Wheat Improvement Center's Insect-Resistant Maize for Africa project has primarily followed this model by first leveraging the private sector as a source of project funding and by later transferring events from the private to the public sector. In contrast, the East Coast Fever vaccine development project involved a higher level of research engagement between the principal organizations—the International Livestock Research Institute and Merial—which can be described as a form of frontier research.

PROJECT COSTS AND BENEFITS

Ideally, the purpose of a PPP-based research project is to lower the costs of research relative to the potential benefits by synergizing both public and private assets. Although ex ante cost–benefit analyses were not conducted or published for many of the PPP projects identified in the 2006 IFPRI study, findings do suggest that for all agbiotech PPPs (including the two cases from Africa), the projects allowed public researchers to conduct research that would have been prohibitively costly had the public sector been working in isolation. The PPP approach allowed public researchers to access financial resources and useful technologies from the private sector that would have been otherwise unavailable in the public domain.

However, findings also suggest that the coordination costs associated with the partnerships—the costs incurred in searching for partners, maintaining partner commitment, and resolving conflicts among partners—were nontrivial. Although it is difficult to quantify the costs of PPP coordination, findings strongly suggest that within-partnership coordination costs are a major challenge to successful PPPs. Findings suggest that the two agbiotech PPPs in Africa were no exception to this.

RISK AND RISK MANAGEMENT

Research projects, whether conducted by the public or private sector, are often risky ventures. In the context of agbiotech, these risks may relate to the general environment in which research is conducted, for example, disruptions caused by negative public opinions on GE crops. Or the risks may be specific to the project and related to the probability that the research process will not yield a successful output or product, will not yield a success within a time horizon that encourages continued investment, or will not yield a product that can pass through legal and regulatory hurdles associated with moving from proof of concept to commercial deployment.

Findings from the 2006 IFPRI study suggest that the agbiotech PPPs are poorly equipped to manage risks associated with the project, whether financial, reputational, or otherwise. This observation is particularly relevant for the two projects in Africa, both of which were initially unable to develop viable products after 5–10 years of research and had to undergo significant changes in project design to continue the research. The financial and reputational risks of limited research outcomes were likely significant in such instances, although efforts to mitigate them have been fairly successful in these examples.[5]

SAFE STEWARDSHIP

Related to the issue of risk is that of safe stewardship of proprietary technologies and materials used in agbiotech research. Findings from the 2006 IFPRI study suggest that safe stewardship may be *the* key issue for private-sector partners looking to engage with public research organizations in agbiotech research in and for developing countries. Concerns revolve around the legal, financial, and reputational risks that could result from the misuse or abuse of agbiotech tools or materials from partners in public research organizations or by third parties that gain unsanctioned access to these tools and materials.

Findings suggest that legal and contractual strategies—indemnifications and disclaimers, for instance—offer private firms some degree of protection. The African Agricultural Technology Foundation, for example, is mandated to facilitate the transfer of technologies (including, but not limited to,

---

5 The Insect-Resistant Maize for Africa project is now working with new transgenic events and additional funding, while the East Coast Fever vaccine development project is being taken up by GALVmed, an international research consortium on livestock. See CIMMYT (2008) on maize and Spielman (2009) on the vaccine.

agbiotech) between research organizations in the public and private sectors (AATF 2009). In doing so, the African Agricultural Technology Foundation provides the expertise—individuals with significant experience in agricultural science, communications, legal affairs, and regulatory affairs—needed to design and negotiate formal agreements that address the risks associated with PPPs. But the findings indicate that this is not perceived as adequate protection against risk for private firms.

Regardless of how skilled such organizations as the African Agricultural Technology Foundation are at mitigating risk through legal recourse, or how well a CGIAR center's own legal capabilities are developed, they are still likely to be limited relative to those of the multinational firms with which they partner. Thus, several respondents to the 2006 IFPRI study argued the need to bolster legal expertise at the system- or center-level sufficiently for the CGIAR to confidentially navigate protracted litigation or negotiate with batteries of lawyers from the private sector. Most other respondents, however, thought that legal recourse offered little benefit to any of the parties to a partnership, arguing that legal recourse would only lead to costly litigation and the loss of good faith among partners, thus harming project implementation and the long-term growth of PPPs. Moreover, many argued that legal recourse is difficult to pursue in developing countries, where legal and regulatory regimes are rarely equipped to address the complex issues underlying PPPs and technology development. Ultimately, respondents indicated that well-planned and carefully executed projects were the only real defense against the risks associated with ensuring good stewardship.

Policy Recommendations

Evidence from the two studies examined here indicate that even though public research in Africa is advancing in several countries, policies may be hindering the advancement of this research. Regulatory processes are holding up testing and commercialization, and institutional and organizational barriers to PPPs are inhibiting the application of private-sector resources and expertise that would provide valuable learning and information-exchange opportunities. These findings suggest that existing policies are insufficient relative to the requirements needed to realize the benefits of these new technological opportunities.

There are several regional, national, and global policy options that could improve agbiotech and GE crop research in Africa. One is to enhance the quantity and quality of information on the environmental safety of GE crops in confined testing or commercial use through information sharing among countries and researchers—such as information about the characteristics of transgenes, gene constructs, host plants, agroecological and agroclimatic zones, experimental designs and observations, and regulatory findings.[6]

An option is to place this information in an open-access venue, such as the Biosafety Clearing-House, so that environmental assessments of crops or traits can be carried out based on accumulated experience among industrialized and developing countries.[7] This approach presents opportunities for South-South collaboration, information networking, and data sharing, with the objective of minimizing redundancies while maximizing the flow of information and expertise based on solid and comprehensive sources of information, ultimately increasing regulatory proficiency and minimizing R&D costs. Greater knowledge of the array of available transgenes can also be used to strengthen the public sector's position in negotiating access agreements over proprietary materials and techniques. In this context, there is also a need to build capacity in research organizations and train local researchers to make effective use of electronic biotechnology research databases and conduct advanced research.

Several innovative approaches to collaborative research could also improve the pace and level of research on agbiotech and GE crops. One possibility is for the public sector to take a stronger public negotiating stance, advocate for greater private tax incentives, or promote other mechanisms to improve the willingness of firms to invest in or provide intellectual property donations for research with a public-interest focus. Other arrangements may be formalized as commercial joint ventures, in which public and private collaborators establish a legal entity to execute a public-interest research agenda and endow it with a mix of governance and management characteristics from the public and private sectors. Lessons can be learned from China, where several agbiotech ventures are advancing as commercial entities spun off from public research agencies, often wholly or majority owned by the parent agency.

Alternatively, researchers and policymakers may explore the use of "honest brokers" (nonprofit third-party organizations) to facilitate interactions

---

6 The Program for Biosafety Systems operates one granting program for research in these areas—the Biotechnology–Biodiversity Interface. Grants are awarded annually following peer review. See PBS (2009).

7 The Biosafety Clearing-House is a mechanism set up by the Cartagena Protocol on Biosafety to facilitate the exchange of information on living modified organisms and assist parties to the Protocol in meeting their obligations under it. The Clearing House provides access to scientific, technical, environmental, legal, and capacity-building information in all six of the United Nations' official languages. See BCH (2009).

between the sectors, manage the research, and assume responsibility for the use of proprietary knowledge and technology. The International Service for the Acquisition of Agri-biotech Applications (ISAAA) and the African Agricultural Technology Foundation are playing such a role in agbiotech and GE crop research in Africa.

The advancement of agbiotech and GE crop research in Africa also requires greater investment in building systems and markets for seed and planting materials. Enactment of plant variety rights and truth-in-labeling laws, combined with a greater commitment from public research organizations to moving technologies off the shelf and into farmers' fields, would facilitate greater investment in GE research and product deployment in Africa. PPPs, technology commercialization programs, competitive grants, reward/prize programs, and other such approaches could go a long way toward shifting public research incentives toward more commercially viable outcomes.

Conclusions

Progress in agbiotech and GE crop research, development, and dissemination in Africa is constrained by insufficient investment in—and regulatory impediments to—the approval and release of new GE products. The two studies on agricultural research in developing countries examined in this chapter offer several critical findings about this progress. First, agbiotech and GE crop research is advancing slowly, although there are some signs that new crops, traits, and technologies are in the pipeline for Africa. Second, although some progress has been made in terms of introducing biosafety regulation in many African countries, movement through regulatory processes is inadequate relative to the opportunities offered by the new technologies. Third, critical assets and competencies from the private sector are not being adequately brought to bear on the research challenge and are not in close collaboration with public research.

These conclusions strongly suggest that efforts need to be redoubled to promote research, development, and dissemination of GE crops in Africa. For agbiotech to benefit Africa, greater efforts are needed to enhance the international exchange of information on GE crops and to overcome institutional barriers to research collaboration between the public and private sectors. Such efforts would promote a more entrepreneurial culture of innovation and make public research institutions and private companies more responsive to emerging needs and opportunities.

Success may depend on the emergence of a real breakthrough—the successful navigation through regulatory processes and deployment through commercial channels of a crop that can make a real difference to small-scale, resource-poor farmers. Such a breakthrough could demonstrate the technology's potential to contribute to the region's development, as well as the importance of the processes needed to make this contribution. However, if the impediments discussed in this chapter persist, the pace of research, development, and dissemination will be insufficient to generate such a breakthrough, thus slowing the diffusion of new technological opportunities and the potential gains to social and economic welfare in Africa.

References

AATF (African Agricultural Technology Foundation). 2009. *AATF Projects*. Accessed November 22. www.aatf-africa.org/projects.php.

Aerni, P. 2006. "Mobilizing Science and Technology for Development: The Case of the Cassava Biotechnology Network (CBN)." *AgBioForum* 9 (1): 1–14.

Agbios. 2009. *Home Page and GM Database Search Page*. Accessed November 22. http://agbios .com/main.php and http://agbios.com/dbase.php.

Alhassan, W. S. 2003. *Agrobiotechnology Application in West and Central Africa: 2002 Survey Outcome*. Ibadan, Nigeria: International Institute of Tropical Agriculture.

Atanassov, A., A. Bahieldin, J. Brink, M. Burachik, J. I. Cohen, V. Dhawan, R. V. Ebora, et al. 2004. *To Reach the Poor—Results from the ISNAR–IFPRI Next Harvest Study on Genetically Modified Crops, Public Research, and Policy Implications*. Environment and Production Technology Division Discussion Paper 116. Washington, DC: International Food Policy Research Institute.

Baum, M., A. de Kathen, and J. Ryan, eds. 2001. *Developing and Harmonizing Biosafety Regulations for Countries in West Asia and North Africa*. Aleppo, Syria: International Center for Agricultural Research in the Dry Areas.

BCH (Biosafety Clearing-House). 2009. *Biosafety Clearing-House*. Accessed November 22. http://bch.cbd.int/.

Beintema, N. M., and G. J. Stads. 2006. *Agricultural R&D in Sub-Saharan Africa: An Era of Stagnation*. Agricultural Science and Technology Indicators Initiative Background Report. Washington, DC: International Food Policy Research Institute.

———. 2011. *African Agricultural R&D in the New Millennium: Progress for Some, Challenges for Many*. Washington, DC: International Food Policy Research Institute.

Birner, R., and N. Linacre. 2008. *Regional Biotechnology Regulations: Design Options and Implications for Good Governance.* IFPRI Discussion Paper 00753. Washington, DC: International Food Policy Research Institute.

Byerlee, D., and K. Fischer. 2001. "Accessing Modern Science: Policy and Institutional Options in Developing Countries." *IP Strategy Today* 1: 1–27.

CERA (Center for Environmental Risk Assessment). 2011. *GM Crop Database.* Accessed September 8. http://cera-gmc.org/index.php?action=gm_crop_database.

Chataway, J. 2005. "Introduction: Is It Possible to Create Pro-Poor Agriculture-Related Biotechnology?" *Journal of International Development* 17 (5): 597–610.

CIMMYT (International Maize and Wheat Improvement Center). 2008. Project: *Insect Resistant Maize for Africa (IRMA).* Accessed November 22. www.cimmyt.org/english/wpp/gen_res /irma.htm.

Cohen, J. I. 2005. "Poor Nations Turn to Publicly Developed GM Crops." *Nature Biotechnology* 23: 27–33.

Cohen, J. I., and R. Paarlberg. 2004. "Unlocking Crop Biotechnology in Developing Countries— A Report from the Field." *World Development* 32: 1563–1577.

Dubock, A. C. 2003. "Learning from Public–Private Partnerships for GM Crops." Paper presented at the 7th ICABR International Conference on Public Goods and Public Policy for Agricultural Biotechnology, June 29–July 3, in Ravello, Italy.

Falck-Zepeda, J., J. I. Cohen, J. Komen, and P. Zambrano. 2003. "Advancing Public Sector Bio-technology in Developing Countries: Results from the Next Harvest Study." Paper presented at the 7th ICABR International Conference on Public Goods and Public Policy for Agricultural Biotechnology, June 29–July 3, in Ravello, Italy.

FAO (Food and Agriculture Organization of the United Nations). 2011. *Database of Biotechnologies in Use in Developing Countries (FAO-BioDeC).* Accessed November 14, 2011. www.fao.org /biotech/inventory_admin/dep/default.asp.

Gisselquist, D., and C. Van der Meer. 2001. *Regulations for Seed and Fertilizer Markets: A Good Practice Guide for Policymakers.* Rural Development Working Paper 22817. Washington, DC: World Bank. www-wds.worldbank.org/servlet/WDSContentServer/WDSP/IB/2001/10/05 /000094946_01092504010357/Rendered/PDF/multi0page.pdf.

Hall, A. 2005. "Capacity Development for Agricultural Biotechnology in Developing Countries: An Innovation Systems View of What It Is and How to Develop It." *Journal of International Development* 17 (5): 611–630.

Harsh, M. 2005. "Formal and Informal Governance of Agricultural Biotechnology in Kenya: Participation and Accountability in Controversy Surrounding the Draft Biosafety Bill." *Journal of International Development* 17 (5): 661–677.

ISAAA (International Service for the Acquisition of Agri-Biotech Applications). 2009. *The International Service for the Acquisition of Agri-Biotech Applications (ISAAA).* Accessed November 22. www.isaaa.org/.

James, C. 2011. *Global Status of Commercialized Biotech/GM Crops: 2011.* ISAAA Brief 43. Ithaca, NY, US: International Service for the Acquisition of Agri-Biotech Applications.

Johanson, A., and C. Ives. 2001. *An Inventory of Agricultural Biotechnology for the Eastern and Central African Region.* East Lansing, MI, US: Michigan State University, Agricultural Biotechnology Support Project.

Karembu, M., F. Nguthi, and I. Abdel-Hamid. 2009. *Biotech Crops in Africa: The Final Frontier.* Nairobi: ISAAA AfriCenter.

Komen, J., H. Webber, and J. Mignouna. 2000. *Biotechnology in African Agricultural Research: Opportunities for Donor Organizations.* ISNAR Briefing Paper 43. The Hague, the Netherlands: International Service for National Agricultural Research.

Mugabe, J. 2002. *Biotechnology in Sub-Saharan Africa: Towards a Policy Research Agenda.* Nairobi, Kenya: Technology Policy Studies Network.

Paarlberg, R. 2001. *The Politics of Precaution—Genetically Modified Crops in Developing Countries.* Baltimore, MD, US: Johns Hopkins University Press.

Pardey, P. G., N. M. Beintema, S. Dehmer, and S. Wood. 2006. *Agricultural Research: A Growing Global Divide?* Food Policy Report 17. Washington, DC: International Food Policy Research Institute.

PBS (Program for Biosafety Systems). 2009. *Program for Biosafety Systems.* Accessed November 22. http://programs.ifpri.org/pbs/.

Pingali, P. L., and G. Traxler. 2002. "Changing Locus of Agricultural Research: Will the Poor Benefit from Biotechnology and Privatization Trends?" *Food Policy* 27: 223–238.

Pray, C. E. 2001. "Public–Private Sector Linkages in Research and Development: Biotechnology and the Seed Industry in Brazil, China and India." *American Journal of Agricultural Economics* 83 (3): 742–747.

Pray, C. E., and A. Naseem. 2003. *The Economics of Agricultural Biotechnology Research.* Agricultural Development Economics Division Working Paper 03-07. Rome: Food and Agriculture Organization of the United Nations.

Pray, C., P. Bengali, and B. Ramaswami. 2005. "The Cost of Biosafety Regulations: The Indian Experience." *Quarterly Journal of International Agriculture* 44: 267–289.

Sithole-Niang, I., J. I. Cohen, and P. Zambrano. 2004. "Putting GM Technologies to Work: Public Research Pipelines in Selected African Countries." *African Journal of Biotechnology* 3 (11): 564–571.

Smale, M., P. Zambrano, G. Gruère, J. Falck-Zepeda, I. Matuschke, D. Horna, L. Nagarajan, et al. 2009. *Measuring the Economic Impacts of Transgenic Crops in Developing Agriculture during the First Decade: Approaches, Findings, and Future Directions.* Food Policy Review 10. Washington, DC: International Food Policy Research Institute.

Smith, J. 2005. "Context-Bound Knowledge Production, Capacity Building and New Product Networks." *Journal of International Development* 17 (5): 647–659.

Spielman, D. J. 2007. "Pro-Poor Agricultural Biotechnology: Can the International Research System Deliver the Goods?" *Food Policy* 32 (2): 189–204.

———. 2009. "Public–Private Partnerships and Pro-Poor Livestock Research: The Search for an East Coast Fever Vaccine." In *Enhancing the Effectiveness of Sustainability Partnerships: Summary of a Workshop*, edited by D. Vollmer, 99–102. Washington, DC: National Academies Press.

Spielman, D. J., and K. von Grebmer. 2004. *Public–Private Partnerships in Agricultural Research: An Analysis of Challenges Facing Industry and the Consultative Group on International Agricultural Research.* Environment and Production Technology Division Discussion Paper 113. Washington, DC: International Food Policy Research Institute.

Spielman, D. J., J. I. Cohen, and P. Zambrano. 2006a. "Policy, Investment, and Partnerships for Agricultural Biotechnology Research in Africa: Emerging Evidence." *African Technology Development Forum Journal* 3 (4): 3–11.

———. 2006b. "Will Agbiotech Applications Reach Marginalized Farmers? Evidence from Developing Countries." *AgBioForum* 9 (1): 23–30.

Spielman, D. J., F. Hartwich, and K. von Grebmer. 2007. *Sharing Science, Building Bridges, and Enhancing Impact: Public–Private Partnerships in the CGIAR.* IFPRI Discussion Paper 708. Washington, DC: International Food Policy Institute.

———. 2010a. "Public–Private Partnerships and Developing-Country Agriculture: Evidence from the International Agricultural Research System." *Public Administration and Development* 30 (4): 261–276.

———. 2010b. "Agricultural Research, Public–Private Partnerships, and Risk Management: Evidence from the International Agricultural Research System." *Asian Biotechnology and Development Review* 12 (1): 21–50.

Tripp, R. 2000. *Strategies for Seed System Development in Sub-Saharan Africa: A Study of Kenya, Malawi, Zambia and Zimbabwe.* Working Papers Series 2. Bulawayo, Zimbabwe: International Crops Research Institute for the Semi-Arid Tropics.

———. 2001. "Can Biotechnology Reach the Poor? The Adequacy of Information and Seed Delivery." *Food Policy* 26: 249–264.

Tripp, R., and N. Louwaars. 1997. "Seed Regulation: Choices on the Road to Reform." *Food Policy* 22: 433–446.

Wafula, D., and N. Clark. 2005. "Science and Governance of Modern Biotechnology in Sub-Saharan Africa—The Case of Uganda." *Journal of International Development* 17 (5): 679–694.

Yerramareddy, I., and P. Zambrano. 2011. *bEcon: Economics Literature about the Impacts of Genetically Engineered Crops in Developing Economies.* Accessed September 9, 2012. www.mendeley.com/groups/1296883/becon/.

Chapter 8

Genetically Modified Foods and Crops: Africa's Choice

Robert Paarlberg

The future of genetically modified (GM) foods and crops in Africa will depend heavily on choices African governments make regarding the regulation of this technology. There are two different regulatory approaches to choose between: the approach used by the European Union (EU) and that used by the United States. There are four key differences between these approaches:

- The regulatory approach used in Europe requires new and separate laws that are specific to GM foods and crops. In contrast, the United States regulates genetically modified organisms (GMOs) for food safety and environmental safety using the laws that were already in place to govern non-GM foods and crops.

- The European approach also requires the creation of new institutions (for example, national biosafety committees) and a separate screening and approval process for GMOs. In the United States the institutions that screen and approve GMOs (the Food and Drug Administration, the Animal and Plant Health Inspection Service, and the Environmental Protection Agency) are the same institutions that screen and approve non-GM foods and crops.

- The European approach also differs because it can decline to approve a new technology on grounds of "uncertainty" alone, without any evidence of risk. A hypothetical risk that has not yet been tested for is sufficient reason for blockage. This is known as the precautionary approach. In the United States, if standard tests for known risks (such as toxicity, allergenicity, and digestibility) have been passed successfully, there is usually no regulatory barrier to commercial release.

- Finally, in Europe products in the marketplace with some GMO content must carry identifying labels, whereas in the United States the Food and Drug Administration does not require labels on any approved GM foods.

Which of these two approaches is better? In the abstract, the best regulatory approach is one that allows new technologies to be used while preventing new risks to human health or the environment. Using this standard, the US approach has so far done a better job than the European approach, because it has allowed many more useful new technologies to be employed by farmers, fortunately without any documented new risks. In contrast, the European approach has blocked the planting of GM crops in most countries in Europe, and in many cases the consumption of GM foods and feeds, to the frustration of most European farmers, who want to share in the productivity gains these crops provide. Most of the GMOs that have been put on the market over the past dozen years have been approved using the risk-based American regulatory approach rather than the uncertainty-based European regulatory approach, and yet the safety record for the technology has remained essentially unblemished. This could be seen as a strong recommendation for the American approach. If the European approach had been followed everywhere, many fewer productive technologies would have been available to farmers, and the safety record would not have been any better.

There has not yet been any documented evidence that approved GMOs have posed new risks either to human health or to the environment. This finding of "no new risks" is now the official view of scientific authorities in Europe itself. European science academies took a number of years to study the impacts of GM crops on human health and the environment following the first commercializations in 1995, but by 2001–04 a consensus had emerged, even in Europe, that no new risks from these seeds had been documented.

In 2001 the Research Directorate General of the EU released a summary of 81 separate scientific studies conducted over a 15-year period (all financed by the EU rather than private industry) aimed at determining whether GM products were unsafe, insufficiently tested, or underregulated (Kessler and Economidis 2001). The EU Research Directorate concluded from this study that "research on GM plants and derived products so far developed and marketed, following usual risk assessment procedures, has not shown any new risks on human health or the environment" (EU Directorate-General for Research and Innovation, press briefing, 2001).

National academies of science in Europe began drawing this same conclusion one year later. In December 2002, the French Academy of Sciences (2002,

p. xxxviii) stated that "all the criticisms against GMOs can be set aside based for the most part on strictly scientific criteria." At the same time the French Academy of Medicine (2002) announced it had found no evidence of health problems in the countries where GMOs had been widely eaten for several years. In the UK in May 2003, the Royal Society presented to a government-sponsored review two submissions that found no credible evidence that GM foods were more harmful than non-GM foods, and the Vice-President and Biological Secretary of the Royal Society, Professor Patrick Bateson, expressed irritation at the undocumented assertions of risk that continued to come from anti-GMO advocates:

> We conducted a major review of the evidence about GM plants and human health last year, and we have not seen any evidence since then that changes our original conclusions. If credible evidence does exist that GM foods are more harmful to people than non-GM foods, we should like to know why it has not been made public. [Paarlberg 2008]

In March 2004, the British Medical Association, which had earlier withheld judgment, endorsed these Royal Society conclusions (BMA 2004). In September 2004 the Union of the German Academies of Science and Humanities produced a report that concluded, "according to present scientific knowledge it is most unlikely that the consumption of the well characterized transgenic DNA from approved GMO food harbors any recognizable health risk" (Helt 2004, 4). This report added that food from insect-resistant GM maize was probably healthier than from non-GM maize due to lower average levels of the fungal toxins that insect damage can cause.

A consensus also emerged at the global scientific level that no new risks had been linked to any of the GM crops and foods to have reached the market so far. In March 2000 the Organisation for Economic Co-operation and Development in Paris organized a conference with 400 expert participants from a variety of backgrounds. These experts announced their agreement that "no peer-reviewed scientific article has yet appeared which reports adverse effects on human health as a consequence of eating GM food" (OECD 2000, 2). In August 2002 the Director-General of the World Health Organization endorsed consumption of GM foods, saying, "[the World Health Organization] is not aware of scientifically documented cases in which the consumption of these foods has negative human health effects. These foods may therefore be eaten" (Mantell 2002).

Some accept that GM foods are probably safe to eat yet still question their safety for other living things in the biological environment (their "biosafety").

All farming disturbs and changes nature, so it is difficult to agree on exactly what level of disturbance can be considered acceptable. For example, planting a GM variety of beet or rapeseed can help farmers control weeds in the field (compared to conventional beet or rapeseed), and as a result there may be fewer insects in the farm field (using the weeds for food and shelter) and fewer weed seeds for some farmland birds to eat. Some might see this as a damaging disturbance of nature. Yet by most conventional definitions of biosafety, the GM crops currently on the market have not disturbed nature (beyond farm fields) any more than conventional crops do. A 2003 study conducted by scientists from New Zealand and the Netherlands published in *The Plant Journal* examined data collected worldwide up to that time, and the authors concluded from this data that the GM crops approved so far had been no more likely to worsen weed problems than are conventional crops, no more invasive or persistent, and no more likely to lead to gene transfer. There was no evidence that GM crops had transferred to other organisms (including weeds) new advantages, such as resistance to pests or diseases or tolerance to environmental stress (Connor, Glare, and Nap 2003).

Later in 2003 the International Council for Science examined the findings of roughly 50 different scientific studies that had been published in 2002–03 and concluded, "there is no evidence of any deleterious environmental effects having occurred from the trait/species combinations currently available" (International Council for Science 2003, 3). In May 2004 the Food and Agriculture Organization of the United Nations issued a 106-page report summarizing evidence that "to date, no verifiable untoward toxic or nutritionally deleterious effects resulting from the consumption of foods derived from genetically modified foods have been discovered anywhere in the world" (FAO 2004). On the matter of environmental safety, this FAO report found the environmental effects of the GM crops approved so far—including such effects as gene transfer to other crops and wild relatives, weediness, and unintended adverse effects on non-target species (such as butterflies)—had been similar to those that already existed from conventional agricultural crops. Finally, in 2007 a study done for the journal *Advances in Biochemical Engineering/Biotechnology* surveyed 10 years of research published in peer-reviewed scientific journals, scientific books, reports from regions with extensive GM cultivation, and reports from international governmental organizations and found that the data available so far provide no scientific evidence that the cultivation of the presently commercialized GM crops has caused environmental harm (Sanvido, Romeis, and Bigler 2007).

In 2010, the EU Directorate-General for Research and Innovation (2010, 16) produced yet another reassuring report on GMO safety:

> The main conclusion to be drawn from the efforts of more than
> 130 research projects, covering a period of more than 25 years of
> research, and involving more than 500 independent research groups, is
> that biotechnology, and in particular GMOs, are not *per se* more risky
> than e.g. conventional plant breeding technologies.

Skeptics who remain fearful sometimes respond that "absence of evidence is not the same thing as evidence of absence." Yet if you look for something for 15 years and fail to find it, that must surely be accepted as evidence of absence. It may not be *proof* that risks are absent, but proving something is absent (proving a negative) is known to be logically impossible.

The explanation for Europe's highly precautionary regulatory approach toward GMOs goes beyond risks. It is a policy posture that reflects not a presence of new risks for Europeans, but instead an absence, for most Europeans, of new benefits. The first generation of GM crops provided benefits to farmers, but almost no benefit at all to food consumers.

The first generation of GM crops that came to the market in 1995–96 provided benefits mostly to farmers growing cotton, maize, and soybeans in the form of lower costs for the control of insects and weeds. Yet Europe does not have many cotton, maize, and soybean farmers, so the new technology had few champions. For the 99 percent of Europeans who were not maize, cotton, or soybean farmers, the new technology offered almost no direct benefit at all. For consumers in Europe, the new GM products did not taste any better, look any better, smell any better, prepare any better, or deliver any improved nutrition. Because the vast majority of Europeans saw little or no direct benefit from the technology, they felt they had nothing to lose by keeping it out of farm fields and out of their food supply. They welcomed a highly precautionary regulatory approach as one way to ensure that outcome.

To demonstrate that it was a benefit calculation rather than a risk calculation that mattered most to Europeans in this case, look at the quite different way Europe regulates GMOs in medicine versus GMOs in agriculture. In the case of medical drugs, Europe does not hesitate to permit the commercial sale of medicines developed with genetic engineering. By 2006 the European Medicines Agency had actually approved 87 recombinant drugs, derived from GM bacteria or from the ovary cells of GM Chinese hamsters. Significantly, these drugs were not free from new risks; it had been learned from clinical

trials that many of these drugs actually increased risks of heart disease, malig-
nancy, and gastric illness, but European regulators approved them just the
same, because of the benefits the drugs could deliver to so many Europeans.
While fewer than 1 percent of Europeans stood to benefit directly from GM
agricultural crops, 100 percent were vulnerable to the diseases these GM drugs
could help treat, so the regulator treatment of the GM drugs was far less precautionary. There were both known risks from clinical trials and plenty of
uncertainties surrounding long-term exposures, yet these risks and uncertainties were not allowed to block the commercial release of a technology that
could bring significant benefits to Europeans.

Consider now the very different circumstances of Africa. In Africa, the percentage of the population that might benefit directly from agricultural GMOs
is much higher than in Europe, because 60 percent or more of all Africans are
still farmers who depend directly on agriculture for income and subsistence.
Some GM crop traits now widely commercialized outside of Africa, such as
crops with the *Bt* gene inserted (for example, maize and cotton), which resist
insect damage with fewer chemical sprays, could have wide benefits if planted
in Africa today. Other GM traits soon to come out of the research pipeline,
including abiotic stress tolerance traits, such as drought resistance, could provide even wider benefits in the future.

Drought-tolerant maize is only one of the new GM crop technologies
now emerging from the research pipeline. Maize is a staple food for more than
300 million people in Africa south of the Sahara (SSA), many of whom are
themselves growers of maize. These Africans remain poor and food insecure
because the productivity of their farming labor is so low. Population growth
has been pushing maize production into marginal areas with little and unreliable rainfall, and only 4 percent of cropland in SSA is irrigated. These factors, combined with human-induced climate change, are expected to increase
drought risks to maize growers in Africa in the years ahead. The development
of maize varieties better able to tolerate drought is one important response to
this growing challenge.

Not all drought-tolerant maize varieties are GMOs. The International
Maize and Wheat Improvement Center's Drought-Tolerant Maize for Africa
initiative, funded in 2007 by the Bill & Melinda Gates Foundation and
the Howard G. Buffet Foundation, is designed to accelerate the breeding
of non-GM drought-tolerant varieties of maize (both hybrids and open pollinated varieties) in 13 countries in SSA. This initiative will use conventional
and marker-assisted selection breeding but no transgenic techniques. A second
initiative does use GM techniques. This is the Water-Efficient Maize for Africa

(WEMA) project, funded in 2008 by the Bill & Melinda Gates Foundation and operated in Africa by the African Agricultural Technology Foundation. The International Maize and Wheat Improvement Center is a partner in this project, as is the Monsanto Company. This initiative will use transgenic techniques in addition to conventional and marker-assisted selection.

Regulatory requirements in Africa for GMOs emerge as a critical consideration here. WEMA's GM varieties of drought-tolerant maize will deliver benefits to African farmers only if African regulators first allow the technology to be tested in open field trials in Africa and then approve the technology for commercial release to farmers. The regulatory gauntlet for this technology will be long and difficult because in Africa, just as in Europe, transgenic technologies are screened using separate and much higher regulatory standards than are used for other technologies. In each separate African country, it will not be possible for technology developers such as the African Agricultural Technology Foundation to conduct research on a WEMA variety (for example, plant a field trial) without an explicit prior approval from a National Biosafety Committee (NBC). Giving or selling the seed to farmers will not be permitted in any country until the NBC has granted a formal commercial release.

Before they grant a commercial release, NBCs typically require that technology developers submit a substantial dossier of data—including the molecular characterization of the variety, the results of lab tests for food safety, and the results of field trials for efficacy and biosafety. Once this data is in hand, the NBC can either grant a commercial release promptly; refuse to approve; ask for more data; or do nothing at all, in which case the technology cannot be legally sold or distributed to farmers. So far, only two governments in SSA have ever given a commercial release to any GM crops: South Africa (for maize, soybean, and cotton) and Burkina Faso (only for cotton) (James 2011).

Even if there are no arbitrary regulatory slowdowns or blockages, it will still require years for GM varieties of drought-tolerant tropical white maize to make their way through this regulatory gauntlet in Africa. The first year of WEMA field trials of GM white maize hybrids was completed in 2009 at two sites in South Africa. These were subtropical varieties intended to be used by smallholders in South Africa and parts of Mozambique. Later in 2009 national biosafety regulators in both Kenya and Uganda followed South Africa and approved applications to begin confined field trials of WEMA maize, but regulators in Tanzania did not. Moving beyond the confined field trial stage will be a challenge for Uganda, because the parliament there has not yet passed a national biosafety bill, a measure that GM critics insist is needed before a full environmental release of any GMO is permitted. In Kenya, moving beyond the confined field trial stage for WEMA maize will be difficult as well. Field trials of GM cotton have been underway in Kenya for years, yet as of 2011 approval for commercial planting had yet to be given. Meanwhile, not even confined field trials have been approved by Tanzania or Mozambique, so even if everything works perfectly for the technology, 2013 is now the earliest that field testing of the WEMA varieties can be undertaken in all five WEMA countries, and it will not be until 2015, at the earliest, that WEMA's GM drought-tolerant tropical white maize hybrids will have undergone enough efficacy testing, agronomic trials, biosafety testing, and variety development in these countries to generate the data needed to support an application to an NBC for commercial release. Even at this point, there will be little guarantee of a prompt regulatory approval.

Why have so many governments in Africa chosen to follow this highly precautionary European approach toward regulating GM foods and crops, despite the technology blockages and extended delays nearly certain to result? Five separate channels of external influence on Africa have led to this choice of Europe's regulatory approach over the approach of the United States.

Bilateral foreign assistance is the first channel of external influence on Africa. Governments in Africa are still significantly dependent on foreign assistance, on average, four times as aid-dependent relative to gross domestic product as the rest of the developing world. For this reason, much that takes place in Africa today remains donor driven. Because Africa's official development assistance from Europe is three times as large as that from the United States, it is the voice of European donors in Africa that tends to be more dominant than any American voice. Governments in Europe have used their official development assistance to encourage African governments to draft and implement European-style regulatory systems for agricultural GMOs.

A second channel of external influence has been multilateral technical assistance through the United Nations Environment Programme (UNEP) / Global Environment Facility (GEF) Global Project for Development of National Biosafety Frameworks. Of 23 African governments that had completed a National Biosafety Framework under this UNEP program by October 2006, all but South Africa and Zimbabwe had no previous regulations in place for agricultural GMOs, so UNEP was in effect writing on a blank slate. In the end, 21 of these 23 countries embraced the strongest possible approach (the "Level One" approach), requiring regulations through binding legal instruments approved by the legislative branch of government (parliament), parallel to the European approach. Europe had greater influence than the United States over this UNEP/GEF program because European governments contribute roughly three times as much to the GEF trust fund as does the United States.

A third channel of external influence has been advocacy campaigns against GMOs from international nongovernmental organizations, the most active of which are headquartered in Europe. Greenpeace International and Friends of the Earth International, both based in Amsterdam, have campaigned heavily in Africa against agricultural GMOs. Zambian officials were told by Greenpeace that if GMOs were let into their country, organic produce sales to Europe would collapse. An organization named Genetic Food Alert warned Zambia in 2002 of the "unknown and unassessed implications" of eating GM foods, and a British group named Farming and Livestock Concern warned them that GM corn could form a retrovirus similar to HIV. These assertions, which were not backed by any evidence, frightened the Zambians into banning GMOs completely.

A group of mostly European nongovernmental organizations continued this campaign against GMOs at the 2002 World Summit on Sustainable Development in Johannesburg, South Africa. Led by Friends of the Earth International, they coached their African partners into signing an open letter warning that GMOs might cause allergies, chronic toxic effects, and cancers. At this same meeting in 2002, two Dutch organizations, HIVOS and NOVIB, joined with partner groups from Belgium, Germany, and the UK to finance a "small-farmers march" on Johannesburg (led by a non-farmer) that ended with a pronouncement that Africans "say NO to genetically modified foods."

A fourth channel of external influence has been commercial agricultural trade. Africa's farm exports to Europe are six times as large as exports to the United States, so it is European consumer tastes and European regulatory systems that Africans most often must adjust to. In 2000 private European buyers stopped importing beef from Namibia because it had been fed on GM maize from South Africa, and then in 2002 Zambia rejected GM maize as food aid in part because an export company (Agriflora) and the export-oriented Zambia National Farmers Union were anxious that exports of organic baby corn to Europe not be compromised. The risks of export rejections from African countries that plant GMOs are actually quite small, as evidenced by the continued growth of food sales to Europe from South Africa, yet anxieties surrounding export loss play a political role in setting policy.

The final channel of external influence is cultural. Most policymaking elites in Africa have much closer cultural ties to Europe than to the United States, so they are naturally inclined to view European practices as the best practices. For example, the Kenyan author of a 2004 article (published by the European-financed nongovernmental organization Participatory Ecological Land Use Management Association) that was titled "Twelve Reasons for Africa to Reject GM Crops," later explained to a newspaper reporter, "Europe has more knowledge, education. So why are they refusing [GM foods]? That is the question everybody is asking" (Paarlberg 2008, 145). Policymaking elites in Africa have often been educated in Europe, they send their children to European schools, and they travel to Europe frequently both on official and unofficial business. It is not surprising that they would be inclined to adopt European-style regulations for GMOs, even though Africa's needs and circumstances are so different from those of Europe.

External influence of this kind is not unique to Africa, of course. In Latin America, within the sphere of influence of the United States, government policies toward GM crops have usually been closer to the American approach than to the European approach. As of 2008, 7 out of the top 10 countries around the world with significant plantings of GMOs were located in the Western Hemisphere. It is also telling that the only Asian country to have approved GMO maize, the Philippines, is a former American colony.

In this case political leaders in Africa pay a price for simply "doing what Europeans do." Europe has placed stifling regulations on GM foods and crops because Europe itself has little need for this new technology. European farmers are already highly productive without it, and European consumers are already well fed. Indeed, like consumers in the United States, Europeans are increasingly overfed. In Africa, where farmers are not yet productive and where so many consumers are not yet well fed, the potential gains that GM crops can provide are more costly to do without.

Rather than deferring to outsiders, either Europeans or Americans, Africans might usefully look for ways to make independent judgments of their own regarding how to regulate GM crops. Other countries in the developing world that still have large farming sectors and operate relatively free from external influence—such as the People's Republic of China—have so far seen high value in this technology and have been investing significant public budget resources of their own to develop this technology, for their own distinct and independent benefit.

References

BMA (British Medical Association). 2004. "Genetically Modified Foods and Health: A Second Interim Statement." March. London.

Connor, A. J., T. R. Glare, and J.-P. Nap. 2003. "The Release of Genetically Modified Crops into the Environment. Part II. Overview of Ecological Risk Assessment." *Plant Journal* 33: 19–46.

EU Directorate-General for Research and Innovation. 2010. *A Decade of EU-Funded GMO Research (2001–2010)*. EUR 24473 EN. Brussels.

FAO (Food and Agriculture Organization of the United Nations). 2004. *State of Food and Agriculture 2003–04: Agricultural Biotechnology: Meeting the Needs of the Poor?* Section B, Part 5. Rome.

French Academy of Medicine. 2002. "OGM et sante." Recommendations (Alain Rerat). Communiqué adopted on December 10. Paris.

French Academy of Sciences. 2002. "Genetically Modified Plants." Report on Science and Technology 13 (December). Paris.

Helt, H. W. 2004. *Are There Hazards for the Consumer When Eating Food from Genetically Modified Plants?* Union of the German Academies of Science and Humanities, Commission on Green Biotechnology. Göttingen: Universität Göttingen.

International Council for Science. 2003. *New Genetics, Food and Agriculture: Scientific Discoveries—Societal Dilemmas*. www.icsu.org.

James, C. 2011. *Global Status of Commercialized Biotech/GM Crops: 2011*. ISAAA Brief 43. ISAAA, Ithaca, NY, US: International Service for the Acquisition of Agri-Biotech Applications.

Kessler, C., and I. Economidis, ed. 2001. *EC-Sponsored Research on Safety of Genetically Modified Organisms: A Review of Results*. Luxembourg: Office for Official Publications of the European Communities.

Mantell, K. 2002. "WHO Urges Africa to Accept GM Food Aid." Science and Development Network, August 30. Accessed January 2012. www.scidev.net/News.

OECD (Organisation for Economic Co-operation and Development). 2000. "GM Food Safety: Facts, Uncertainties, and Assessment, Rapporteurs' Summary." Presented at the OECD Conference on the Scientific and Health Aspects of Genetically Modified Foods, February 28–March 1, in Edinburgh.

Paarlberg, R. 2008. *Starved for Science: How Biotechnology Is Being Kept out of Africa*. Cambridge, MA, US: Harvard University Press.

Sanvido, O., J. Romeis, and F. Bigler. 2007. "Ecological Impacts of Genetically Modified Crops: Ten Years of Field Research and Commercial Cultivation." *Advanced Biochemical Engineering/Biotechnology* 107: 235–278.

Conclusion

Guillaume Gruère, Idah Sithole-Niang, and José Falck-Zepeda

A lthough genetically modified (GM) crops have been adopted by farmers in an increasing number of countries, their use remains very limited in African countries south of the Sahara. Only three countries have approved GM crops for commercialization. South Africa first planted GM crops in 1997; Burkina Faso and Egypt just started using GM maize and cotton in 2008 and 2009, respectively. Other African countries south of the Sahara have implemented confined field trials of GM crops (James 2011), but so far none of these crops has reached farmers' fields. Detractors have used the observed limited adoption to support the allegation that GM crops are not—and will not be—useful for African countries south of the Sahara. Even though this argument is deceptively simplistic, it rightly reflects the wider questioning about the current and potential role GM crops and products could play in the region in a changing global economic, demographic, and climatic environment.

This book offers a collection of economic and policy studies providing some elements of response to this particular question. Although there is still a significant gap in the literature on the economic effects of GM crops in those African countries south of the Sahara, especially compared to other regions, the selected contributions presented here aim to show that existing research can already bring forward useful lessons for stakeholders and policymakers in this area while pointing toward areas for further inquiries.

More specifically, as noted in the following two sections, we have identified five main lessons from the results of the contributed chapters that outline both the opportunities offered by GM crops and the multiple challenges ahead. Naturally, these conclusions are directly drawn from the individual studies with specific geographic scopes, topics, and methods that can present important caveats. But the following sections demonstrate that they are representative of other publications on GM crops in those African countries south of the Sahara. Thus, despite the heterogeneous set of countries, crops, and methodologies discussed, these lessons convey a message that so far is validated by available research-based evidence.

Opportunities for African Farmers South of the Sahara

The first main lesson is that, *based on available data and published studies, current GM crops have had on average a positive economic effect in African countries south of the Sahara, but the magnitude and distribution of their potential economic benefits for farmers highly depend on the crop, trait, and especially the institutional setting in which the technology is introduced.* An important corollary to this finding is the need for ex ante technology assessment studies to go beyond simple cost–benefit analyses using average performance data and to address such issues as uncertainty, downside risk, and production practices and their limitations. This finding is consistent with the conclusions of Smale et al. (2009) based on an international review of published articles on GM crops in developing economies. It also concurs with the main conclusions of Tripp (2009) based on a compilation of case studies on the use of Bt (insect-resistant) cotton in different countries (including South Africa and others in Asia and Latin America) and with meta-analyses conducted by Finger et al. (2011) and Areal, Riesgo, and Rodriguez-Cerezo (2012).

In a review of the situation and of published economic studies, Gouse (Chapter 1) indicates that the adoption of Bt crops by smallholder farmers in South Africa appears to have resulted in positive agricultural and economic outcomes for a majority of adopters. GM crops in general have been extensively adopted there, with annual fluctuations. The South African example of Bt cotton also shows the importance of a proper institutional setting.

Pray et al. (Chapter 2) suggest that the adoption of Bt maize could reduce the exposure of poor farmers in South Africa to the mycotoxin fumonisin and therefore lower the risk of certain types of cancers. But a wider adoption of Bt maize by smallholder farmers in South Africa would not solve the problems of poverty and exposure, and the authors note that Bt maize would need to be adopted in challenging areas where farmers are not currently using GM or hybrid crops.

Two ex ante analyses in this book conclude that GM crops have a significant role to play in Uganda and thus potentially in other African countries south of the Sahara. Horna et al. (Chapter 3) show that GM cotton has the potential to improve the productivity of cotton in Uganda. At the same time the authors conclude that it will not be a silver bullet for resolving the poor performance of cotton in the country, especially given the lack of adequate input. Even though their simulation results demonstrate that the highest returns are associated with Bt and herbicide-tolerant cotton, they argue that these technologies will not increase the profitability of cotton very much, as existing low productivity in the cotton sector is a binding factor limiting potential improvements from a damage-control technology such as Bt cotton. Institutional and regulatory issues will need to be addressed if these GM crops are to be successfully commercialized.

Kikulwe et al. (Chapter 4) analyze the case of GM bananas in Uganda using survey data. The authors estimate the annual opportunity cost of non adoption of these crops as an average 38 US dollars (US$) per Ugandan household. This means that, if the approval decision were made solely on the basis of economic benefits, GM bananas should be immediately released unless the average household is willing to pay US$38 per year to avoid its release, production, and consumption. Furthermore, Uganda loses a significant amount of money by not adopting because of the potentially catastrophic damage that the fungal disease black Sigatoka causes on resource-poor farmers who do not have the resources for chemical control.

These results are in line with other economic evaluations of GM crops in other African countries south of the Sahara. Several ex ante simulation studies on Bt cotton in West Africa have concluded that the region would benefit from releasing this technology and would lose from avoiding it (Vitale et al. 2007; Falck-Zepeda, Horna, and Smale 2008; Bouët and Gruère 2011). Burkina Faso has adopted GM cotton, but it is too early to quantify its full economic effect in the field, even if results from the field trials (Vitale et al. 2008) and first observations suggest a positive economic effect with yield increases and pesticide reduction (Fasozine 2009; Vitale et al. 2010). An ex ante study of GM vegetables, based on farmer surveys in Ghana, also showed that even if the results depend on the crop, GM vegetables are likely to be economically advantageous for farmers in Ghana (Horna et al. 2008). Several studies have also focused on Bt cowpeas, with converging evidence on the potential for using GM technologies to improve this widely used staple crop in West Africa (Langyintuo and Lowenberg-DeBoer 2006; Gbègbèlègbè et al. 2009).

Challenges Ahead

Various challenges have constrained the use of GM crop technologies. The studies presented in this book analyze some of the key issues that need to be addressed, from the development of adapted GM technologies to the setting up of regulations and the handling of regulatory burdens and trade and market acceptance.

The second main lesson is that *there are insufficient efforts in public and private biotechnology development in Africa, and that one of the main constraints is related to the policy environment.* Spielman and Zambrano (Chapter 7) argue that although public research is advancing, with promising technologies for the African context (such as drought-tolerant maize), policy environments may hinder progress in deploying such technologies. The authors find that the private sector is still left out of the research environment in agricultural biotechnology in Africa and note the need to promote research and development, increase exchanges of information, and promote a more entrepreneurial culture of innovation.

The third lesson is that *evolving biosafety regulations in countries south of the Sahara, which tend to determine the degree of deployment of GM crops in the region, appear to be based on a highly costly, European precautionary approach, despite clearly diverging agricultural and development priorities.* Paarlberg (Chapter 8) demonstrates that external influence has a critical role in Africa's choice of regulatory model. Yet the author argues that the European model does not correspond to African needs and realities for a number of reasons, and he concludes that African regulators will be better off by choosing their policies independently of others. Falck-Zepeda and Zambrano (Chapter 6) review biosafety regulatory costs for applicants, showing the high costs for selected East African projects. The authors argue that high regulatory costs can have serious consequences in African biotechnology development. They conclude that biosafety regulatory systems need to balance risk avoidance with the cost of implementation and the potential (net) benefits that the technology may bring when adopted by farmers.

The fourth lesson is that the *alleged short-term export risks due to potential market losses in Europe and other GM-averse countries may have been exaggerated and need to be assessed on a case-by-case basis,* and that *the upcoming challenges of market access and import regulations call for regional integration of GM trade regulations.*

Wafula and Gruère (Chapter 5) discuss the potential export risks for southern and eastern African countries based on a review of results from the literature. Although each study the authors review presents intrinsic limitations, their findings indicate that commercial risk—and possible export loss to the EU—appear to be relatively small in the short and medium terms. Intraregional trade appears to be more important, and therefore regional agreements are critical to the future of GM crops in Africa. Their conclusions are nuanced by the fact that trade may change in the future.

Other trade analyses in the literature confirm these conclusions, but also note the cost of not adopting GE crops when competitors do. For instance, in the case of export risks, Anderson and Jackson (2005) show that the welfare loss with European market restrictions is small compared to the expected gains with GM crops in African countries south of the Sahara. The Langyintuo and Lowenberg-DeBoer (2006) analysis of the West African regional market for cowpeas concludes that regional introduction of Bt cowpeas would be the best solution to prevent nonadopting countries in the region from losing economically. Similarly, at the global level, the delayed adoption of Bt cotton in West and Central Africa is found to have a significant cost, given the adoption of this technology by competitors (for example, see Elbehri and MacDonald 2004). More generally, Gruère, Bouët, and Mevel (2011) draw three lessons from the international trade literature on GM crops that do apply to African countries south of the Sahara: (1) GM crop adoption generates economic gains for adopting countries and importing nonadopters, (2) domestic regulations can reduce these gains, and (3) import regulations in other countries can also affect the gains of exporting adopters.

One aspect not discussed in this book is the design and implementation of import regulations in African countries. Kagundu (2009) provides a first assessment of border-control challenges for GM products in Kenya. Kimani and Gruère (2010) discuss the proposed cost of case-by-case import approval and the possible implementation of documentation requirements under the Cartagena Protocol in the same country. Gruère and Sengupta (2010) mention the incomplete enforcement of border measures in southern Africa. More studies are needed in this area.

The last lesson is that *the level of awareness of GM crops appears to be low among surveyed consumers.* In addition, acknowledging this low awareness and the limitation it may confer on survey results, and the fact that only one study is included here, *GM technology seems to be generally well accepted among surveyed consumers, but urban (especially high-income) consumers appear to have a lower acceptance of GM food than do rural consumers.* If confirmed, this low acceptance may create significant challenges on the road to commercialization of potentially promising GM crops, especially foodcrops and those crops developed by public-sector research.

As mentioned earlier, Kikulwe et al. (Chapter 4) study the acceptance of GM bananas in Uganda. Among other things, the authors find that urban elites are more reluctant to buy GM products than are rural consumers. Thus, developers and national policymakers need to implement explicit robust outreach and communication efforts to address public concerns and raise awareness in terms of the potential impact of the introduction of GM crop technologies.

There are a few other studies published on this topic for the region, and they confirm this result. Kimenju and De Groote (2008) analyzed GM maize acceptance in Kenya and show that awareness of biotechnology is limited in Kenya, especially in rural areas. All consumers are willing to buy GM maize at the same price as non-GM maize, even if there are some concerns. Environmental concerns are found only among urban consumers. In their study of Bt cowpeas in Benin, Niger, and Nigeria, Gbègbèlègbè et al. (2009) found that rural consumers, who tend to be farmers, would be willing to pay more for Bt cowpeas than for the non-GM varieties. In contrast, urban consumers are less willing to buy GM cowpeas. In South Africa, a study run by the Ministry of Science and Technology on consumers' perceptions of biotechnology found that 75 percent of the respondents were uninformed about biotechnology (Durham 2009).

A relevant side issue to the policy debate in African countries south of the Sahara in the literature is the possible potential disconnect between consumer perceptions and the local food industry, which may lead to conflicts between these groups, highlighting the need for improved knowledge-exchange channels. Bett, Okuro Ouma, and de Groote (2010) clearly illustrate the latter point in their study of Kenyan gatekeepers' perspectives, noting that these companies would benefit if they were informed about the results of consumer surveys. They also provide evidence of the disconnect between consumers and the local industry. Kenyan gatekeepers (millers and supermarkets) are found to be more skeptical than consumers are on the possible use of GM food. Most would prefer assessing GM food on a case-by-case basis before purchasing or selling these products. For some of the larger commercial actors, this perception may be related to the great reluctance of buyers abroad regarding the use of GM food. For instance, Gruère and Sengupta (2009) found that GM-free private standards set up by large importing companies in European and other developed countries do influence traders' decisions and indirectly some policy decisions about biotechnology in eastern and southern African countries. Gruère and Takeshima (2012) provide an economic analysis explaining why importing companies can indeed affect views of decisionmakers in Africa.

Outlook for the Future

These conclusions confirm that GM crops have a significant role to play in agriculture development in African countries south of the Sahara. But the nature and modalities of this role will be defined largely by national and regional policy choices as well as the institutional setting in which these technologies will be deployed to farmers. Although there is evidence that several GM crops could be beneficial to smallholder farmers in Africa, on a case-by-case basis, there is also much work needed to improve the policy, research, and regulatory environment to ensure that they can be a tool to help advance agricultural development. The research results collected in this book suggest that setting up a balanced and functioning regulatory framework is a critical and necessary determinant of the advancement of GM crops. Given current export and import issues, regulating trade of GM products will require regional coordination if not a common framework. In parallel, governments should increase the awareness of farmers, consumers, and the food sector concerning GM crops.

A side result of this book is the call for additional research to be conducted in this area. More data and more rigorous analyses are needed to confirm or complement the lessons of these selected contributions. Only a few countries and crops have been studied, and there are only a few consumer studies in the peer-reviewed published literature. But even focusing on the specific studies presented in this book, there are still important limitations to the research methodologies. This is certainly not limited to studies in Africa (Smale et al. 2009), but this point needs to be emphasized. Each conclusion naturally stands on the robustness of the arguments, data, and methods used in each particular study.

Other issues that have not been specifically addressed in this book require a lot more research. First, the role of intellectual property rights in facilitating or constraining adoption will continue to be part of the discussions. Second, the acceptance and impacts of GM crops on women farmers, given their importance in African farming and agricultural development, is another priority area for future research. A few projects are ongoing in Africa, but more will need to be done, especially as new GM technologies advance toward commercialization. Third, the potential environmental benefits and costs associated with GM technologies in Africa and other regions is still an underinvestigated topic. Although GM crops are often associated with potentially negative environmental externalities, there is increasing evidence that some of them may also have positive environmental externalities (for example, the effect of Bt crop adoption on water quality or non-target organisms, and broader protection of biodiversity by increasing agricultural productivity). Last, more analyses of the political economic rationale driving different countries' positions for or against GM crop introduction are needed to understand some of the above-listed challenges.

What is the outlook for the future? The development or testing of current and new GM crops and the strengthening of regulatory capacity in several African countries suggest that GM crops will play a growing role in the agriculture of African countries south of the Sahara. But their impact on smallholder farmers is uncertain and will depend on the issues outlined in this book.

Clearly, next-generation GM crops—such as those tolerant to drought, resistant to pest and diseases, and with improved nutritional quality—could address other more specific productivity constraints that African farmers face every day. Other examples include crops with increased efficiency in the use of macronutrients, such as nitrogen or phosphorus. Furthermore, they could contribute to managing risk and address the expected increase in climate change variability. They would thus support African communities' resiliency and improve the livelihoods of resource-poor farmers.

But this potential can only be made possible through robust and consistent investments in technology and the support of the institutional setting where these technologies will be released to farmers. The driving force for these technologies to advance lies in the hands of national policymakers and their willingness to pursue these opportunities. But the capacity, expertise, and support of external institutions will also continue to play major roles. In particular, new players like China and the Bill & Melinda Gates Foundation acting as donors and technology providers, and such regional bodies as the African Union and the Common Market for Eastern and Southern Africa, can be expected to play a critical role in determining the future of these agricultural technologies. Given the evidence presented here, undertaking these challenges proactively appears to be the best approach for the future.

References

Anderson, K., and L. A. Jackson. 2005. "Some Implications of GM Food Technology Policies for Sub-Saharan Africa." *Journal of African Economies* 14 (3): 385–410.

Areal, F. J., L. Riesgo, and E. Rodriguez-Cerezo. 2012. "Economic and Agronomic Impact of Commercialized GM Crops: A Meta-Analysis." *Journal of Agricultural Science.* CJO2012 doi: 10.1017/S0021859612000111.

Bett, C., J. Okuro Ouma, and H. de Groote. 2010. "Perspectives of Gatekeepers in the Kenyan Food Industry towards Genetically Modified Food." *Food Policy* 35 (4): 332–340.

Bouët, A., and G. Gruère. 2011. "Refining Estimates of the Opportunity Cost of Non-adoption of Bt Cotton: The Case of Seven Countries in Sub-Saharan Africa." *Applied Economic Perspectives and Policy* 33 (2): 260–279.

Durham, B. 2009. "Public Perception of Biotechnology in South Africa." Presentation at the International Food Policy Research Institute conference Delivering Agricultural Biotechnology to African Farmers: Linking Economic Research to Decision-Making, May 2009, in Entebbe, Uganda.

Elbehri, A., and S. MacDonald. 2004. "Estimating the Impact of Transgenic Bt cotton on West and Central Africa: A General Equilibrium Approach." *World Development* 32 (12). 2049–2064.

Falck-Zepeda, J. B., D. Horna, and M. Smale. 2008. "Betting on Cotton: Potential Payoffs and Economic Risks of Adopting Transgenic Cotton in West Africa." *African Journal of Agricultural and Resource Economics* 2 (2): 188–207.

Fasozine. 2009. "Le coton transgénique, la voie de salut de la SOFITEX." *Fasozine*, November 20. www.fasozine.com/index.php/component/content/article/109-publireportage/1704-le-coton-transgenique-la-voie-de-salut-de-la-sofitex.

Finger, R., N. El Benni, T. Kaphengst, C. Evans, S. Herbert, B. Lehmann, S. Morse, and N. Stupak. 2011. "A Meta Analysis on Farm-Level Costs and Benefits of GM Crops." *Sustainability* 3: 743–762. doi:10.3390/su3050743.

Gbègbèlègbè, D. S., J. Lowenberg-DeBoer, R. Adeoti, J. Lusk, and O. Coulibaly. 2009. "The Estimated ex ante Economic Impact of Bt Cowpea in Niger, Benin and Northern Nigeria." In *Harnessing the Potential of Agricultural Biotechnology for Food Security and Socio-Economic Development in Africa*. Proceedings of the 1st All Africa Congress on Biotechnology, 378–384. Nairobi, Kenya: Agricultural Biotechnology Stakeholders Forum. Accessed January 30, 2013. www.absfafrica.org/downloads/Congress%20_Proceedings_Publication.pdf.

Gruère, G. P., and D. Sengupta. 2009. "The Effects of GM-Free Private Standards on Biosafety Policymaking in Developing Countries." *Food Policy* 34 (5): 399–406.

———. 2010. "Reviewing South Africa's Marketing and Trade-Related Policies for Genetically Modified Products." *Development Southern Africa* 27 (3): 333–352.

Gruère, G. P., and H. Takeshima. 2012. "Will They Stay or Will They Go? The Political Influence of GM-Averse Importing Companies on Biosafety Decision Makers in Africa." *American Journal of Agricultural Economics* 94 (3): 736–749.

Gruère, G. P., A. Bouët, and S. Mevel. 2011. "International Trade and Welfare Effects of Biotechnology Innovations: GM Food Crops in Bangladesh, India, Indonesia and the Philippines." In *Genetically Modified Food and Global Welfare*, edited by C. A. Carter, G. Moschini, and I. M. Sheldon, 283–308. Frontiers of Economics and Globalization, Volume 10. Bingley, UK: Emerald.

Horna, J. D., M. Smale, R. Al-Hassan, J. Falck-Zepeda, and S. Timpo. 2008. *Insecticide Use on Vegetables in Ghana: Would GM Seed Benefit Farmers?* IFPRI Discussion Paper 785. Washington, DC: International Food Policy Research Institute.

James, C. 2011. *Brief 43: Global Status of Commercialized Biotech/GM Crops: 2011*. Ithaca, NY, US: International Service for the Acquisition of Agri-Biotech Applications.

Kagundu, A. 2009. "Mechanisms for Regulating GM Imports in Africa." Paper presented at the International Food Policy Research Institute conference Delivering GM Technology to African Farmers: Linking Economic Research to Decision-Making, May 19–21, in Entebbe, Uganda.

Kimani, V., and G. Gruère. 2010. "Implications of Import Regulations and Information Requirements under the Cartagena Protocol on Biosafety for GM Commodities in Kenya." *AgBioForum* 13 (3): 222–241.

Kimenju, S., and H. de Groote. 2008. "Consumer Willingness to Pay for Genetically Modified Food in Kenya." *Agricultural Economics* 38 (1): 35–46.

Langyintuo, A. S., and J. Lowenberg-DeBoer. 2006. "Potential Regional Trade Implications of Adopting Bt Cowpea in West and Central Africa." *AgBioForum* 9 (2): 111–120.

Smale, M., P. Zambrano, G. Gruère, J. Falck-Zepeda, I. Matuschke, D. Horna, L. Nagarajan, et al. 2009. *Measuring the Economic Impacts of Transgenic Crops in Developing Agriculture during the First Decade: Approaches, Findings and Future Directions*. Food Policy Review 10. Washington, DC: International Food Policy Research Institute.

Tripp, R., ed. 2009. *Biotechnology and Agricultural Development: Transgenic Cotton, Rural Institutions and Resource-Poor Farmers*. London and New York: Routledge.

Vitale, J., T. Boyer, R. Uaiene, and J. H. Sanders. 2007. "The Economic Impacts of Introducing Bt Technology in Smallholder Cotton Production Systems of West Africa: A Case Study from Mali." *AgBioForum* 10 (2): 71–84.

Vitale, J., H. Glick, J. Greenplate, M. Abdennadher, and O.Traoré. 2008. "Second-Generation Bt Cotton Field Trials in Burkina Faso: Analyzing the Potential Benefits to West African Farmers." *Crop Sciences* 48: 1958–1966.

Vitale, J. D., G. Vognan, M. Ouattarra, and O. Traore. 2010. "The Commercial Application of GMO Crops in Africa: Burkina Faso's Decade of Experience with Bt Cotton." *AgBioForum* 13 (4): 320–332.

Contributors

Ekin Birol (e.birol@cgiar.org) is the head of the Impact and Policy Unit at HarvestPlus Division of the International Food Policy Research Institute. Her related research includes the following co-authored publications: "Bi-modal Preferences for Bt Maize in the Philippines: A Latent Class Model" (*AgBioForum*, 2012), "A Latent Class Approach to Investigating Developing Country Consumers' Demand for Genetically Modified Staple Food Crops: The Case of GM Banana in Uganda" (*Agricultural Economics*, 2011), and "Investigating Heterogeneity in Farmer Preferences for Milpa Diversity and Genetically Modified Maize in Mexico: A Latent Class Approach" (*Environment and Development Economics*, 2009).

José Falck-Zepeda (j.falck-zepeda@cgiar.org) is a senior research fellow in the Environment and Production Technology Division of the International Food Policy Research Institute (IFPRI) and leader of the polity team in IFPRI's Program for Biosafety Systems. His relevant publications include the following co-authored articles: "Estimates and Implications of the Costs of Compliance with Biosafety Regulations in Developing Countries: The Case of the Philippines and Indonesia" (*GM Crops and Food: Biotechnology and Agriculture in the Food Chain*, 2011) and "Socio-Economic Considerations in Biosafety and Biotechnology Decision Making: The Cartagena Protocol and National Biosafety Frameworks" (*Review of Policy Research*, 2011). Falck-Zepeda is the author of "Socio-Economic Considerations, Article 26.1 of the Cartagena Protocol on Biosafety: What Are the Issues and What Is at Stake?" (*AgBioForum*, 2009) and is co-editor of the 2011 special issue of *AgBioForum* "Farmers and Researchers Discovering Biotech Crops: Experiences Measuring Economic Impacts among New Adopters."

Marnus Gouse (Marnus.gouse@up.ac.za) is a postdoctoral fellow in the Department of Agricultural Economics, Extension and Rural Development of the University of Pretoria, South Africa. His related research includes the following co-authored publications: "A GM Subsistence Crop in Africa: The Case of Bt White Maize in South Africa (*International Journal of Biotechnology,* 2005), "GM Maize as Subsistence Crop: The South African Smallholder Experience" (*AgBioForum,* 2012), "Ten Years of Bt Cotton in South Africa: Putting the Smallholder Experience into Context" (*Biotechnology and Agricultural Development: Transgenic Cotton, Rural Institutions and Resource-Poor Farmers,* 2009).

Guillaume Gruère (ggruere@gmail.com) was a senior research fellow in the Environment and Production Technology Division of the International Food Policy Research Institute. His related research includes the following co-authored publications: "Refining Opportunity Cost Estimates of Not Adopting GM Cotton: An Application in Seven Sub-Saharan African Countries (*Applied Economic Perspectives and Policy,* 2011), "GM-Free Private Standards and Their Effects on Biosafety Decision-making in Developing Countries" (*Food Policy,* 2009), "Will They Stay or Will They Go? The Political Influence of GM-Averse Importing Companies on Biosafety Decision Makers in Africa" (*American Journal of Agricultural Economics,* 2012).

Daniela Horna (jdhorna@fastmail.fm) was a postdoctoral fellow in the Environment and Production Technology Division of the International Food Policy Research Institute. She has co-authored "Farmer Willingness to Pay for Seed-Related Information: Rice Varieties in Nigeria and Benin" (*Environment and Development Economics,* 2007), *Insecticide Use on Vegetables in Ghana: Would GM Seed Benefit Farmers?* (IFPRI discussion paper, 2008), and "Betting on Cotton: Potential Payoffs and Economic Risks of Adopting Transgenic Cotton in West Africa" (*African Journal of Agricultural and Resource Economics,* 2008).

Enoch Kikulwe (ekikulw@gwdg.de) is a postdoctoral fellow, Department of Agricultural Economics and Rural Development, Georg-August University of Göttingen, Germany. His related research includes the following co-authored publications: "A Latent Class Approach to Investigating Demand for Genetically Modified Banana in Uganda" (*Agricultural Economics,* 2011), "Attitudes, Perceptions, and Trust: Insights from a Consumer Survey Regarding Genetically Modified Banana in Uganda" (*Appetite,* 2011), and "Rural Consumers' Preferences for Banana Attributes in Uganda: Is There a Market for GM Staples?" (in J. Bennett and E. Birol, eds., *Choice Experiments in Developing Countries: Implementation, Challenges and Policy Implications,* 2010).

Miriam Kyotalimye (m.kyotalimye@asareca.org) is a program assistant at the Association for Strengthening Agricultural Research in Eastern and Central Africa, Entebbe, Uganda.

Robert Paarlberg (rpaarlberg@wellesley.edu) is a professor of Political Science at Wellesley College and adjunct professor of Public Policy at the Harvard Kennedy School in Massachusetts. He is the author of *Food Politics: What Everyone Needs to Know* (2010) and *Starved for Science: How Biotechnology Is Being Kept out of Africa* (2008).

Carl E. Pray (pray@aesop.rutgers.edu) is a professor in the Department of Agriculture, Food and Resource Economics at Rutgers University, United States. Co-authored publications include "Addressing Micronutrient Deficiencies: Alternative Interventions and Technologies" (*AgBioForum,* 2007), "Insect-Resistant GM Rice in Farmers' Fields: Assessing Productivity and Health Effects in China" (*Science,* 2005), and "GM Cotton and Farmers' Health in China: An Econometric Analysis of the Relationship between Pesticide Poisoning and GM Cotton Use in China" (*International Journal of Occupational and Environmental Health,* 2004).

John P. Rheeder (john.rheeder@mrc.ac.za) is a senior scientist in the Programme on Mycotoxins and Experimental Carcinogenesis Unit at the Medical Research Council, Cape Town, South Africa. He has co-authored the following related publications: "Production of Fumonisin Analogs by *Fusarium* species" (*Applied Environmental Microbiology,* 2002), "Toxicity, Pathogenicity and Genetic Differentiation of Five Species of *Fusarium* from Sorghum and Millet" (*Phytopathology,* 2005), and "Guidelines on Mycotoxin Control in South African Foodstuffs: From the Application of the Hazard Analysis and Critical Control Point (HACCP) System to New National Mycotoxin Regulations" (*Medical Research Council Policy Brief,* www.mrc.ac.za, 2009).

Theresa Sengooba (t.sengooba@cgiar.org) is East African Regional Coordinator, Program for Biosafety Systems (PBS), International Food Policy Research Institute, Entebbe, Uganda. She is the co-author of the following relevant publications: *Assessing the Potential Impact of Genetically Modified Cotton in Uganda* (PBS Policy Note, 2009), *Analysis of the Biosafety System in Uganda* (PBS Uganda Country Study, 2005), and "Biosafety Education Relevant to Genetically Engineered Crops for Academic and Non-academic Stakeholders in East Africa" (*Electronic Journal of Biotechnology,* 2009).

Gordon S. Shephard (gordon.shephard@mrc.ac.za) is chief specialist scientist in the Programme on Mycotoxins and Experimental Carcinogenesis Unit of the Medical Research Council, Tygerberg, South Africa. His publications include "Risk Assessment of Aflatoxins in Food in Africa" (*Food Additives and Contaminants*, 2008) and co-authored publications "Fumonisin Mycotoxins in Traditional Xhosa Maize Beer in South Africa" (*Journal of Agriculture and Food Chemistry,* 2005) and "Exposure Assessment for Fumonisins in the Former Transkei Region of South Africa" (*Food Additives and Contaminants,* 2007).

Idah Sithole-Niang (isn@iwayafrica.co.zw) is a molecular biologist at the University of Zimbabwe's Department of Biochemistry, Technical Advisor for south of the Sahara Africa to International Food Policy Research Institute's Program for Biosafety Systems, and is chair of the Board of Trustees of the African Agricultural Technology Foundation. She is author of the following publications: *Regulatory Requirements and Technology Adoption: The Case of Biotech Cotton* (2009), *Transgenic Horticultural Crops on the African Continent* (2011), and *Genetic Engineering for Resistance to Viruses* (2011).

David J. Spielman (d.spielman@cgiar.org) is a senior research fellow in the Environment and Production Technology Division of the International Food Policy Research Institute. His co-authored publications include "Intellectual Property Rights, Private Investment in Research, and Productivity Growth in Indian Agriculture: A Review of Evidence and Options" (*Journal of Agricultural Economics,* 2010), "Public–Private Partnerships and Developing-Country Agriculture: Evidence from the International Agricultural Research System" (*Public Administration and Development,* 2010), and "Private-Sector Investment in R&D: A Review of Policy Options to Promote Its Growth in Developing-Country Agriculture" (*Agribusiness,* 2010).

Yvette Volkwyn [née Yvette Manuel] (yvette.volkwyn@mrc.ac.za) is a senior research technologist in the Programme on Mycotoxins and Experimental Carcinogenesis Unit of the Medical Research Council, Cape Town, South Africa. Her related co-authored publications include "Immunoglobulin Gene Rearrangements of B-cell Non-Hodgkin's Lymphoma Types in Paraffin Embedded Tissue Using PCR" (*European Journal of Haematology*, 1997), "Detection of t(14;18) Translocation in Follicle Center Cell Lymphomas in South African Ethnic Groups Using PCR" (*Leukemia and Lymphoma*, 1998), and "Fumonisin B$_1$ and Risk of Hepatocellular Carcinoma in Two Chinese Cohorts" (*Food and Chemical Toxicology*, 2012).

David Wafula (wafuladavid@yahoo.com) was the Kenya Coordinator for the Program for Biosafety Systems. His publications include "Science and Governance of Modern Biotechnology in Sub-Saharan Africa—The Case of Uganda" (*Journal of International Development*, 2005), "Implications of GMOs on Market Access and Imports in Eastern Africa (*Biotechnology: Eastern African Perspectives on Sustainable Development and Trade Policy*, 2007) and the co-authored "GM Crops and Food: Biotechnology in Agriculture and the Food Chain" (*Landes Bioscience*, 2012).

Justus Wesseler (justus.wesseler@wzw.tum.de) is professor of Agriculture and Food Economics in the Department of Agriculture Economics, Center of Food and Life Sciences Weihenstephan, Technische Universtät München, Germany. His related research includes the following co-authored publications: "The Maximum Incremental Social Tolerable Irreversible Costs (MISTICs) and Other Benefits and Costs of Introducing Transgenic Maize in the EU-15" (*Pedobiologia*, 2007), "Economically Optimal Timing of Crop Disease Control under Uncertainty: An Options Approach" (*Journal of the Royal Society Interface*, 2010), and "Benefits and Costs of Biologically Contained GM Tomatoes and Eggplants in Italy and Spain" (*Sustainability*, 2011).

Liana van der Westhuizen (liana.van.der.westhuizen@mrc.ac.za) is a specialist scientist in the Programme on Mycotoxins and Experimental Carcinogenesis Unit of the Medical Research Council, Tygerberg, South Africa. Her related publications include the co-authored "Biomarkers of Exposure: Mycotoxins— Aflatoxin, Deoxynivalenol and Fumonisins" (in L. Knudsen and D. F. Merlo, eds., *Biomarkers and Human Biomonitoring,* 2012) and "Fumonisin B_1 as a Urinary Biomarker of Exposure in a Maize Intervention Study among South African Subsistence Farmers" (*Cancer Epidemiology Biomarkers and Prevention,* 2011).

Patricia Zambrano (p.zambrano@cgiar.org) is a senior research analyst in the Environment and Production Technology Division of the International Food Policy Research Institute. Her related co-authored publications include "Unweaving the Threads: The Experiences of Female Farmers with Biotech Cotton in Colombia" (*AgBioForum,* 2012), "A Case of Resistance: Herbicide-Tolerant Soybeans in Bolivia" (*AgBioForum,* 2012), and "The Socio-Economic Impact of Transgenic Cotton in Colombia" (*Biotechnology and Agricultural Development: Transgenic Cotton, Rural Institutions and Resource-Poor Farmers,* R. Tripp, 2009).

Index

Page numbers for entries occurring in figures are followed by an *f;* those for entries in notes, by an *n;* and those for entries in tables, by a *t.*